The use of radioactive isotopes in the life sciences

The use of radioactive isotopes in the life sciences

J. M. CHAPMAN and G. AYREY

Queen Elizabeth College,
University of London

London
George Allen & Unwin
Boston Sydney

First published in 1981

GEORGE ALLEN & UNWIN LTD
40 Museum Street, London WC1A 1LU

British Library Cataloguing in Publication Data

Ayrey, G.
The use of radioactive isotopes in life sciences.
1. Radioisotopes in biology
I. Title II. Chapman, J. H.
573'.028 QH324.9.I/

ISBN 0-04-570011-7
ISBN 0-04-570012-5 Pbk

Set in 10 on 11 point Times by Preface Ltd, Salisbury, Wilts.
and printed and bound in Great Britain by
William Clowes (Beccles) Limited, Beccles and London.

Preface

Modern courses in any of the biological sciences would be incomplete without repeated references to the use of radioisotopes. We have been involved for many years in teaching isotope techniques to undergraduate and postgraduate students, and have been aware of the need for a simple text which could be referred to whenever the topic arose in the general biological curriculum. We hope that this present volume fulfils such a need.

Our aim has not been to present the student with a comprehensive treatise but rather with an adequate background knowledge of radioactivity and to outline some of the methods used when working with radioisotopes, including measurement of activity. We have included a chapter on radiation protection, as this is a subject with which many postgraduate students often feel ill at ease, and also a chapter on special analytical methods, some of which are used less than they deserve mainly through ignorance of their existence rather than any inherent practical difficulty in using them.

We have included a few simple experiments at the end of some of the chapters – these illustrate some of the more important aspects of the use of radioisotopes. Readers may find it necessary to adapt some of the experiments to suit their own requirements or to change them so that they will work with the equipment they have available. However, we hope that at least some of them will be attempted, since we believe that the key to a proper grasp of theory depends upon some practical experience, however small. The mathematical content of the various chapters has been reduced to a minimum, but we have attempted to indicate the importance of careful experimental design and the statistical assessment of the validity of experimental data.

We wish to express our gratitude to Professor Leslie Young, who read the entire manuscript and made several helpful observations and to Mrs M. Anderson for much secretarial assistance particularly in typing the manuscript from our illegible first draft.

A book of this size could not include more than a passing reference to the many research workers who have made important contributions to the subject, covering a vast area of scientific literature. However, we freely acknowledge our indebtedness to the published work of generations of physicists, chemists and biologists, and hope that our brief opus will assist those who wish to make a start in, or at least understand some of the intricacies of, this rewarding field of endeavour.

J. M. Chapman
January 1981 G. Ayrey

Acknowledgements

We are grateful to the following for permission to reproduce copyright material:

Professor J. A. Bassham (Fig. 1.1); Professor R. D. Keynes (Fig. 1.3); Pergamon Press (Fig. 2.5); Professor B. W. Fox (Table 4.3); Amersham International Ltd (The Radiochemical Centre) (Table 5.3); Figure 6.8 adapted from *Dating in archaeology* by S. Fleming, by permission of J. M. Dent & Sons Ltd; National Radiological Protection Board (part of Appendix B).

Contents

1 Introduction

1.1 Why use isotopes for biological research?

The availability and use of radioisotopes as tracers for biological work has become of immense importance in many areas of study. The fate of any given compound in an organism or biological process can be traced simply by following the pathway(s) taken by an applied radioisotope; the extent or rate of a process can be measured easily by determining the incorporation of a radioisotope into a particular product; radioisotopes can be used as internal standards for measuring the efficiency of extraction of compounds from biological material; their use in radioimmunoassay allows the determination of extremely small amounts of biologically important compounds. There are literally hundreds of possible applications for isotopes in biology and their use, although requiring special techniques and precautions, has led to the solution of many problems which would otherwise have been very difficult to solve. Some examples are given within this chapter (p. 4).

1.2 Availability and preparation of radioisotopes

The first artificial production of a radioisotope was accomplished in 1934 by Curie and Joliot who succeeded in making phosphorus-30 out of aluminium-27 by bombarding the aluminium with α-particles. Since that time the development of charged particle accelerators (van de Graaff, cyclotrons, etc.) and nuclear reactors has provided the means to produce high energy particles at will and these are now used for the large scale production of an enormous number of different isotopes (about 1500 are known, the majority being artificially produced).

Radioisotopes used in tracer studies have to be produced using nuclear reactions. Suppose we have a substance S which we require to transform into a radioactive substance by bombarding it with high energy particles P. We may write

$$\underset{\text{substance}}{S} + \underset{\text{high energy particles}}{P} \xrightarrow[\text{reaction}]{\text{nuclear}} \underset{\text{radioactive substance}}{R} + \underset{\text{emitted particle}}{EP}$$

In the case of the reaction Curie and Joliot used for the production of phosphorus-30 we may write

$$^{27}\text{Al} + {}^{4}\text{He} \longrightarrow {}^{30}\text{P} + {}^{1}\text{n}$$

α-particle neutron

or, using a shorthand notation,

$$^{27}\text{Al}(\alpha, \text{n})^{30}\text{P}$$

(The superscripts refer to the number of **nucleons** in each element; see p. 12.)

When we write an equation relating to a nuclear reaction involving **nucleons** (see p. 13) we should note that, just as in chemical reactions, matter (and energy, see p. 13) must be conserved. Thus during such a reaction the numbers of protons and neutrons should remain the same. Hence for Curie and Joliot's reaction we can write more fully

$$^{27}_{13}\text{Al} + {}^{4}_{2}\text{He} \longrightarrow {}^{30}_{15}\text{P} + {}^{1}_{0}\text{n}$$

(The subscripts refer to the number of protons in each element; see p. 12.) We can see that there are fifteen protons on the right-hand side of the equation (in the ^{30}P nucleus) and also fifteen on the left-hand side (thirteen in the ^{27}Al nucleus and two in the α-particle). Similarly, the numbers of neutrons on both sides of the equation remain the same.

The most common practical method of isotope production uses neutron irradiation of some material in a nuclear reactor to produce nuclear reactions. In this case the nuclear reactor is acting as a massive source of neutrons, the flux of which can be controlled at will. (A discussion of the working of a nuclear reactor is beyond the scope of this book and the interested student should refer to the reading list in Appendix D.) Table 1.1 shows some examples of the nuclear reactions which are used to produce some of the biologically important radioisotopes.

Table 1.1 Some examples of nuclear reactions which are used to produce biologically important isotopes.

	Target material		*Isotope produced*		*Emitted radiation or particle*
(a)	$^{31}_{15}\text{P} + {}^{1}_{0}\text{n}$	$\longrightarrow$	$^{32}_{15}\text{P}$	+	γ
(b)	$^{32}_{16}\text{S} + {}^{1}_{0}\text{n}$	$\longrightarrow$	$^{32}_{15}\text{P}$	+	$^{1}_{1}\text{p}$
(c)	$^{6}_{3}\text{Li} + {}^{1}_{0}\text{n}$	$\longrightarrow$	$^{3}_{1}\text{H}$	+	$^{4}_{2}\alpha$
(d)	$^{14}_{7}\text{N} + {}^{1}_{0}\text{n}$	$\longrightarrow$	$^{14}_{6}\text{C}$	+	$^{1}_{1}\text{p}$
(e)	$^{35}_{17}\text{Cl} + {}^{1}_{0}\text{n}$	$\longrightarrow$	$^{35}_{16}\text{S}$	+	$^{1}_{1}\text{p}$

Depending upon the sample material in the reaction, many competing nuclear reactions may occur as a result of variations in neutron energies and other factors, therefore the starting material must be carefully selected in order to minimize production of undesirable radioisotopes. In addition, the starting material and its radioproducts must be chemically stable under the conditions in the reactor to avoid unwanted breakdown. Thus, for example, the production of ^{14}C is facilitated by using Be_3N_2 in solid form as the starting material which is stable under the conditions of production. Another problem concerns the purification of the isotope following its production. For example, if ^{32}P is generated from ^{31}P the product is inseparably mixed with unreacted stable ^{31}P, making the specific activity (see p. 21) very low and hence limiting its use as a radiotracer. If, however, the tracer is made by irradiation of sulphur [$^{32}S(n,p)^{32}P$], then the product can be separated from the untreated ^{32}S. In this case the radioisotope represents almost all of the phosphorus in the sample and the product is termed **carrier-free**. Unfortunately the production of carrier-free material is not always possible because a suitable starting material may not exist.

As well as neutron-produced isotopes from reactors many fission products from the reactor itself are useful for biological work (for example ^{90}Sr and ^{131}I).

A second way in which isotopes can be produced is through the use of cyclotrons. The cyclotron is a very versatile tool for isotope production because of the wide variety of type and energies of the accelerated particles it can produce. Because the operating costs of cyclotrons are higher than those of nuclear reactors they are only used for manufacturing isotopes which could not be produced in significant quantities in any other way (e.g. ^{22}Na, ^{26}Al, ^{48}V, ^{54}Mn, ^{74}As and ^{125}I) and also for producing carrier-free material without the need for expensive highly enriched target elements (e.g. for the production of carrier-free ^{55}Fe and ^{85}Sr).

Once the primary radioisotope has been prepared it can be employed directly. Very often, however, in biological research the investigator requires a specifically labelled compound for use in a particular experiment (usually the case for ^{14}C and ^{3}H isotopes). There are many ways of preparing such compounds and the reader should consult one of the appropriate texts listed in Appendix D.

In most cases research workers purchase ready made compounds from specialist firms providing radioisotopes. (A selected list of addresses is given in Appendix E.) Not only do they provide a wide variety of compounds from stock but will also normally consider requests for special custom syntheses. Table 1.2 indicates the various types of labelling which are generally available.

Table 1.2 Various types of labelling available for radioisotopes.

Labelling type	*Description*	*Example*
uniform	label is distributed in a statistically uniform manner throughout the compound	[U-^{14}C]sucrose (produced by photosynthesis using $^{14}CO_2$)
general	label is distributed generally throughout the compound but not uniformly	[G-^{3}H]caffeine
nominal	used where there is some degree of uncertainty as to the position of the label	D,L-[2(n)-^{3}H]glutamic acid (most of label in the '2' position)
specific	used where all of the label is in the position stated	[1-^{14}C]stearic acid

1.3 Examples of the use of radioisotopes in biology

In the rest of this chapter we shall give some examples of how radioisotopes have been used to elucidate biological problems.

(a) The dark reactions of photosynthesis Photosynthesis is a multistep process and the elucidation of the sequence of reactions taking place was the outcome of a series of brilliant experiments initiated by Calvin and his associates at the University of California. Critical to the work was the use of carbon dioxide labelled with ^{14}C. Briefly, the experimental techniques involved growing the green alga *Chlorella* in a culture medium which contained $^{14}CO_2$, stopping the reaction by running the culture into boiling 80% aqueous ethanol and then identifying the compounds into which the carbon dioxide was incorporated. It was found that radioactive carbon turned up in glucose molecules within 30 s after the start of photosynthesis. In order to find out how this occurred it was then necessary to identify the intermediate steps between $^{14}CO_2$ and ^{14}C-labelled glucose. For this purpose the *Chlorella* was exposed to $^{14}CO_2$ for periods of time less than 30 s (in some experiments for less than 1 s!) and the compounds containing ^{14}C separated by chromatography and detected by autoradiography. Such autoradiograms were fairly complex (Fig. 1.1 shows a representation of a typical one) but once the radioactive compounds were separated and located it was then possible to identify them by chemical analysis.

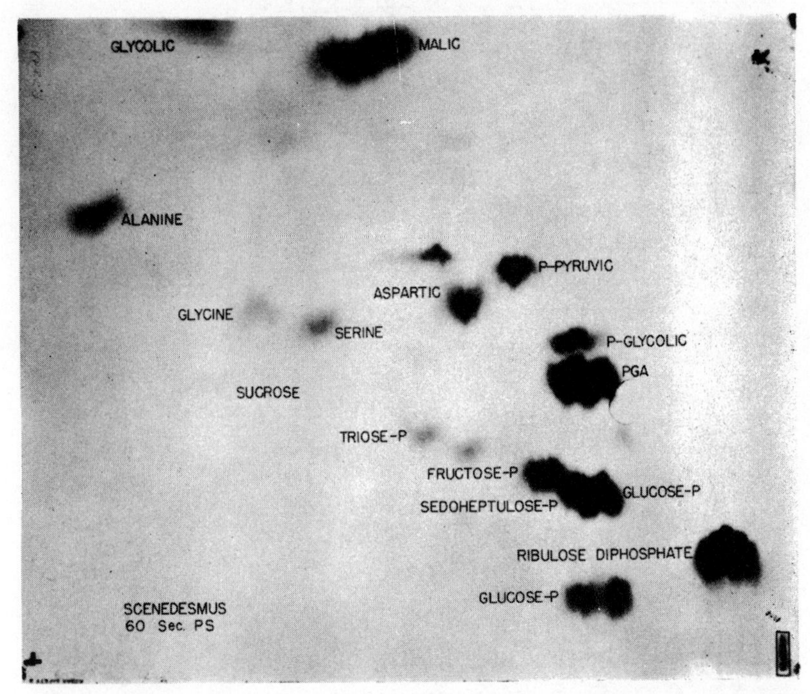

Figure 1.1 Autoradiogram illustrating the products of photosynthesis in a unicellular green algae after exposure to $^{14}CO_2$. (From Bassham, J. A. and M. Calvin 1957. *The path of carbon in photosynthesis,* Englewood Cliffs, NJ: Prentice-Hall.)

Once the various products were identified, the position of the radioactive carbon atoms within them was ascertained by degrading the isolated compounds using standard chemical procedures and locating the radioactivity within the degradation products. In this way, the pattern of carbon dioxide fixation starting with phosphoglyceric acid (PGA) and its transformation products ending with glucose were identified. Figure 1.2 shows a schematic representation of the cycle of dark reactions identifying the possible positions of a ^{14}C atom from $^{14}CO_2$ during one turn of the cycle.

(b) Protein synthesis In the early 1950s biochemists began to use radioactive amino acids to study the processes involved in protein synthesis. Up to this time it was thought that there was very little protein synthesis and breakdown in the cells of adult non-

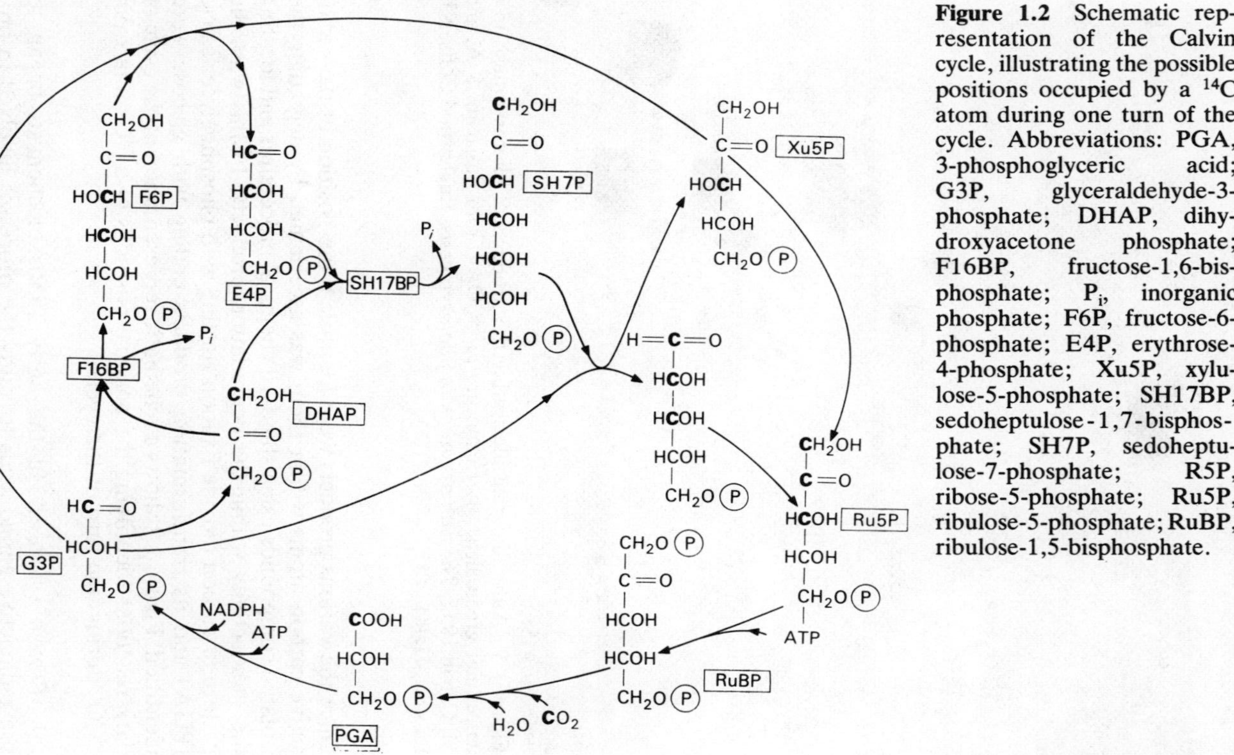

Figure 1.2 Schematic representation of the Calvin cycle, illustrating the possible positions occupied by a ^{14}C atom during one turn of the cycle. Abbreviations: PGA, 3-phosphoglyceric acid; G3P, glyceraldehyde-3-phosphate; DHAP, dihydroxyacetone phosphate; F16BP, fructose-1,6-bisphosphate; P_i, inorganic phosphate; F6P, fructose-6-phosphate; E4P, erythrose-4-phosphate; Xu5P, xylulose-5-phosphate; SH17BP, sedoheptulose-1,7-bisphosphate; SH7P, sedoheptulose-7-phosphate; R5P, ribose-5-phosphate; Ru5P, ribulose-5-phosphate; RuBP, ribulose-1,5-bisphosphate.

growing organisms and it came as a distinct surprise to find that if one injected radioactive amino acids into an animal and then isolated proteins from the various tissues that these proteins were radioactive; albeit some more so than others. Since then our understanding of the mechanism of protein synthesis has increased enormously and many of the critical experiments necessary to our understanding have required the use of (in fact would have been extremely difficult or impossible without) radioisotopes of one sort or another. Some of the most exciting experiments were those involving the synthesis of protein *in vitro* since work of this kind was necessary to establish which cell components were required for protein synthesis. Many of these experiments involved isolating cell components (such as ribosomes) from homogenates of tissues and mixing them together in test tubes with the addition of various other compounds including a radioactive amino acid. After a short period of time any protein present in the mixture was precipitated, usually by the addition of trichloracetic acid. Following this the protein precipitate was subjected to a rigorous washing procedure designed to remove any radioactive amino acid which might be 'sticking' to the protein rather than incorporated within it. Finally, the radioactivity in the precipitate was counted by using a variety of methods, some of which involved dissolving the protein and counting a sample of the solution, and others of which involved counting the precipitate directly on a filter which had been used to

Table 1.3 Typical 'cell-free' mixture for synthesizing proteins.

rabbit reticulocyte lysate (containing ribosomes and soluble cytoplasmic components)
125 mM KCl
1.5 mM $MgCl_2$
1.0 mM ATP (adenosine triphosphate)
an energy regenerating system such as a mixture of creatine phosphate and phosphocreatine kinase
0.2 mM GTP (guanosine triphosphate)
trace amounts of spermine and spermidine
a radioactive amino acid (e.g. L-[^{35}S]methionine)
together with
a mixture of the other 19 commonly occurring amino acids (unlabelled)
a suitable buffer (~pH 7.2)
a preparation containing messenger RNA

collect it. Any radioactivity present in the protein sample was then taken as a measure of the extent of protein synthesis. The contents of a typical 'cell-free' mixture capable of synthesizing protein is shown in Table 1.3. From such early experiments our understanding of protein synthesis has improved to such an extent that it is now possible to purchase kits from various firms for assessing the amount and type of protein synthesis catalysed by different types of messenger RNA, again made possible through the use of radioactive amino acids.

(c) Active transport of nerve impulses Our understanding of the mechanism of nerve impulse transmission has also been assisted by the use of radioisotopes (for example the use of ^{24}Na ions). The current picture of the way in which an impulse is conducted along a nerve fibre is as follows: first a change occurs in the permeability of the nerve cell membrane to sodium ions after stimulation. Sodium ions move into the cell, changing the polarity of the inside from negative to positive (this wave of polarity reversal represents the nerve impulse). Within a few milliseconds the polarity is restored by the outward migration of K^+ ions, and then during a long period of inactivity a 'sodium pump' actively transports sodium ions from the inside to the outside of the nerve cell, causing the cell interior to

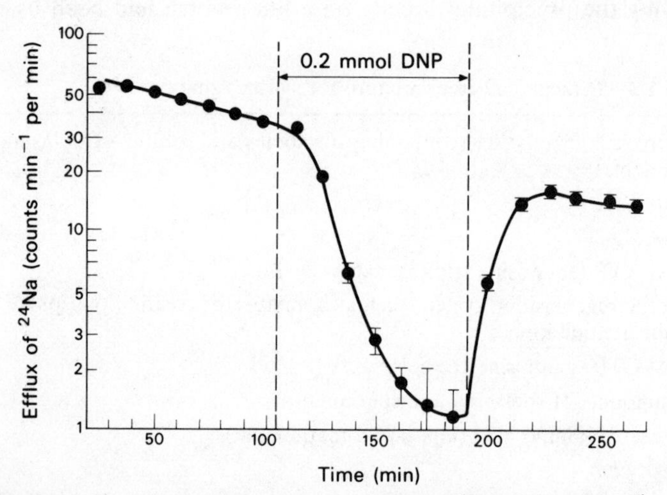

Figure 1.3 The effect of a metabolic inhibitor (2,4-dinitrophenol) on the efflux of radioactive sodium (^{24}Na) from a giant axon of the cuttlefish *Sepia*. (From Hodgkin, A. L. and R. D. Keynes 1955. Active transport of cations in giant axons from *Sepia* and *Loligo*. *J. Physiol., Lond.* **128**, 28–60.)

become highly negative again and allowing the positive potassium ions to move back into the cell, thus restoring the original resting potential.

Figure 1.3 illustrates a typical experiment in which the need for metabolic energy to 'work' the sodium pump is investigated. In this experiment radioactive sodium is used and the efflux of the isotope from a giant axon of *Sepia* (cuttlefish) is monitored with time. Following stimulation the concentration of sodium ions within the cell increases as a consequence of the change in permeability. These ions are then gradually removed as the sodium pump begins to work and this is observed as a fairly high efflux of sodium during the first 100 minutes. The injection of a metabolic poison into the axon stops the pump from working (efflux decreases), but following its removal (by washing out) its activity is restored as evidenced by an increase in the amount of sodium transported out of the cell.

There are numerous examples of the use of radioisotopes in all branches of biological research, some requiring very sophisticated techniques and others, such as in the preceding example, rather more simple ones. The approach used depends almost entirely upon the application. The other chapters in this book are devoted to the background knowledge required for the use of radioisotopes rather than to any one particular example of their use, and it is suggested that before embarking upon any special method which the reader intends to use, the details should be either thoroughly worked out or obtained from the literature.

2 Atoms, isotopes and radioactivity

2.1 Structure of the atom

A modern description of the atom in terms of quantum mechanics is beyond the scope of this book. A pictorial representation, however, using the older Rutherford–Bohr model (Fig. 2.1) is convenient for discussion of the important features. This model visualizes the atom as consisting of a dense central **nucleus** around which **electrons** are moving in fixed orbits which represent fixed energy levels. The orbits are grouped in **shells** which are sequentially filled with specific numbers of electrons as one proceeds along and down the periodic table of elements. Thus we have the K,L,M, . . . shells as shown in Figure 2.1. The outermost electrons are the valency electrons and take part in chemical reactions. If energy is supplied to the atom in some way, electrons will become excited and will jump to higher energy levels. Subsequently, they will return to their normal 'ground state' orbits and the absorbed energy will be re-emitted. If the movements have occurred in the outer electrons, the emitted energy is in the form of light – a good example is the familiar orange sodium street lighting which is due to excitation of sodium atoms by an electric discharge. If larger quantities of energy are supplied it is possible to excite or even eject electrons from the innermost shells, when the subsequent electronic movements give rise to emission of **X-rays** which are electromagnetic radiations of much shorter wavelengths than light (0.01–20 nm).

Rutherford's experiments using α-particles (see p. 17) as projectiles directed at thin metal foils proved that nearly all the mass of the atom was concentrated in a dense central nucleus. We now know that the diameters of atoms are in the range 1×10^{-10}–6×10^{-10} m and that the diameter of the nucleus is approximately 10 000 times less than this. The dimensions are difficult to comprehend, but if we could magnify an atom of carbon until its nucleus was the size of a golf ball (~4 cm) the valency electrons would be orbiting at a distance of 550 m and, although still invisible to the naked eye, would prevent any other atom from approaching any closer to the nucleus because of electrical repulsion. Nevertheless, it is clear that all the solid matter around us consists largely of empty space and that high energy

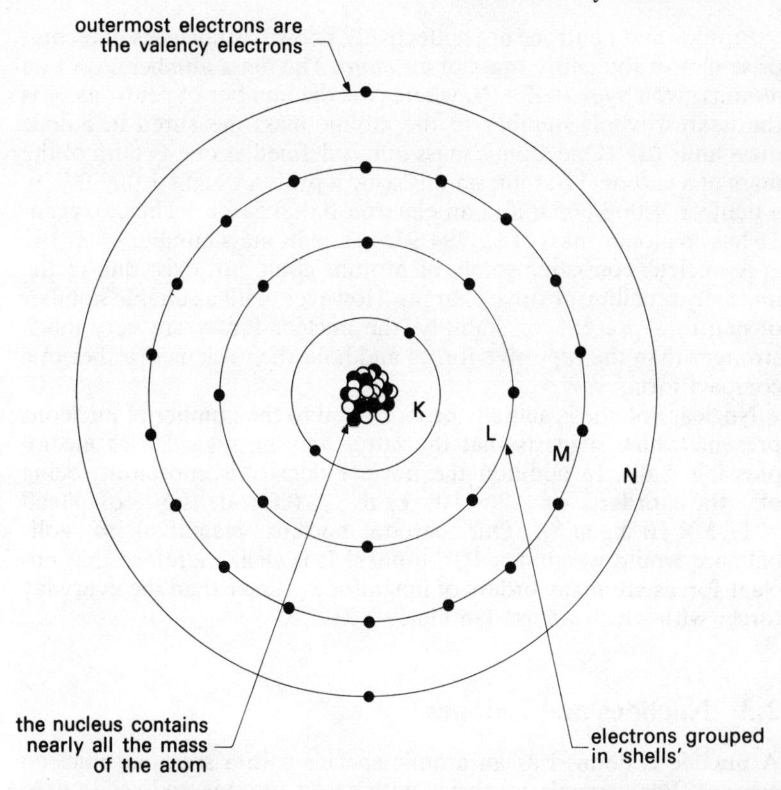

Figure 2.1 Diagrammatic representation of an atom of zinc. The protons and neutrons (nucleons) are grouped together in the nucleus around which the electrons move in orbits which represent fixed energy levels.

nuclear projectiles have no difficulty in passing through for extensive distances.

2.2 Structure of the nucleus

Since atoms are electrically neutral, the nucleus contains the same number of positive charges as electrons in orbit. Each unit positive charge resides on a **proton**, the only charged particle present in the nucleus. The number of protons, Z, is the **atomic number** of the element and determines its position in the periodic table. The other important particle present in the nucleus is the **neutron** which has almost the same mass as a proton but no electric charge.

Protons and neutrons are collectively known as **nucleons** and comprise almost the entire mass of an atom. The **mass number**, A, of an atom is given by $A = Z + N$, where N is the number of neutrons. A is the nearest whole number to the atomic mass measured in **atomic mass units** (u). (One atomic mass unit is defined as one-twelfth of the mass of a carbon-12 atom; on this scale a proton weighs 1.007 277 u, a neutron 1.008 665 u and an electron 0.000 549 u.) Thus, oxygen-16 has an atomic mass of 15.994 915 u and its mass number, A, is 16.

A nucleus consisting solely of protons could not exist due to the mutually repelling positive charges. However, with a suitable number of neutrons present for stability, the nuclear forces are very much stronger than the repulsive forces and hold the nucleus together in a compact form.

Nuclear volume is actually proportional to the number of nucleons present, which suggests that the latter may be regarded as incompressible balls. In addition the nuclear density is enormous, being of the order of 2×10^{17} kg m^{-3} (cf. density of lead $= 1.13 \times 10^{4}$ kg m^{-3}). Our carbon nucleus magnified to golf-ball size would weigh 4×10^{10} tonnes! It is clear therefore that nuclear forces are many orders of magnitude greater than the everyday forces with which we are familiar.

2.3 Nuclides and isotopes

A **nuclide** is defined as an atomic species with a specified nucleon content. For example a species with seven protons and seven neutrons is the nuclide nitrogen-14 whose internationally recognized symbol is $^{14}_{7}N$, although in most cases the Z value in the subscript is superfluous and the symbol used is ^{14}N.

Isotopes are then easily defined as a series of nuclides with the same atomic number but differing mass numbers. Two examples are given in Table 2.1, where the three known isotopes of hydrogen and seven isotopes of carbon are listed. All the elements have isotopes and in some cases large numbers have been identified (e.g. tin has twenty-five isotopes). The important feature worth repeating is that the only difference between isotopes of a given element is in the mass of their nuclei. Z is constant, therefore all the isotopes have the same electronic constitution; their valency electrons are identical so that, chemically, isotopes are very similar. This means that the rare or radioactive species may be used as tracers (see p. 142) for the normal abundant stable isotope. Certain properties which depend upon nuclear mass are dealt with in Chapter 5.

Table 2.1 The isotopes of hydrogen and carbon.

Atomic number, Z	*Neutron number, N*	*Mass number, A*	*Isotope*	*Symbol*	*Half-life*†
1	0	1	protium	^{1}H	stable
1	1	2	deuterium	^{2}H	stable
1	2	3	tritium	^{3}H	12.35 a‡
6	4	10	carbon-10	^{10}C	19 s
6	5	11	carbon-11	^{11}C	20 min
6	6	12	carbon-12	^{12}C	stable
6	7	13	carbon-13	^{13}C	stable
6	8	14	carbon-14	^{14}C	5730 a
6	9	15	carbon-15	^{15}C	2 s
6	10	16	carbon-16	^{16}C	0.7 s

†For a definition of half-life see p. 19.
‡'a' is now accepted as the SI abbreviation for years.

2.4 Stability of the nucleus

In Section 2.2 it was stated that neutrons are present in the nucleus to provide stability. In fact the number of neutrons is quite critical; too few or too many neutrons can cause the nucleus to become unstable. This is clearly illustrated by the isotopes of carbon in Table 2.1. Carbon-12 and carbon-13 are the only stable isotopes which occur in nature with relative abundances of approximately 98.9 and 1.1% respectively. The isotopes of mass number 10 and 11 have too few neutrons and those of mass number 14, 15 and 16 have too many; all are unstable. The unstable nuclei seek to achieve stability by adjusting their balance of protons to neutrons and the processes whereby they achieve this give rise to radioactivity, which is the dissipation of excess energy in the form of ionizing radiations.

2.5 Conservation of mass and energy

When a nucleus emits energy it has to lose an equivalent amount of mass in order to satisfy the physical law of conservation of energy. Einstein's equation expresses the equivalence of mass and energy in the relationship

$$E = mc^2$$

where E is the energy in joules, m is the mass in kilograms and c is the velocity of light in metres per second. In other words mass is regarded as a particular form of energy. Thus,

$$1\text{ g of matter} \equiv 10^{-3} \times (3 \times 10^{8})^{2}\text{ kg m}^{2}\text{ s}^{-2} = 9 \times 10^{13}\text{ J}$$

(which is approximately 25 000 000 kilowatt hours!).

On the atomic scale, energies are conventionally quoted in megaelectronvolts. One electronvolt (1 eV) is the kinetic energy acquired by an electron when it is accelerated through a potential difference of 1 volt in a vacuum:

$$1\text{ eV} = 1.602 \times 10^{-19}\text{ J}$$

Thus,

$$1\text{ u}(=1.660\,43 \times 10^{-27}\text{ kg}) \equiv 1.660\,43 \times 10^{-27} \times (3 \times 10^{8})^{2}\text{ kg m}^{2}\text{ s}^{-2} = 1.49 \times 10^{-10}\text{ J} = 931\text{ MeV}$$

and

$$1\text{ electron }(=0.000\,549\text{ u}) \equiv 0.510\text{ MeV.}$$

2.6 Radioactive disintegration

2.6.1 Neutron-deficient nuclides

(a) Positron emission The carbon isotopes ^{10}C and ^{11}C in Table 2.1 have too few neutrons for stability. To correct this imbalance one of the protons is converted to a neutron and the positive charge is ejected from the nucleus into the surrounding matter in the form of a **positron**, which is represented as a β^+-particle (Fig. 2.2a). The β^+-particle is in fact a positively charged electron, e^+, possessing kinetic energy. A second particle, the **neutrino**, ν, with no charge and negligible mass, is also emitted. Its importance is discussed further in section 2.8 (p. 24). Thus the overall process may be represented as

$$p \rightarrow n + e^{+} + \nu + \text{energy}$$

The positron is an example of an antiparticle which is part of antimatter. It is possible that the universe contains whole galaxies composed of antimatter entirely as stable as our own galaxy. However, in our environment of myriad electrons the positron is rapidly annihilated by collision with an electron when the combined masses are converted to energy (see above) in the form of electromagnetic γ-radiation:

$e^{+} + e^{-} \rightarrow 2 \times 0.51$ MeV γ-radiation **(annihilation radiation)**

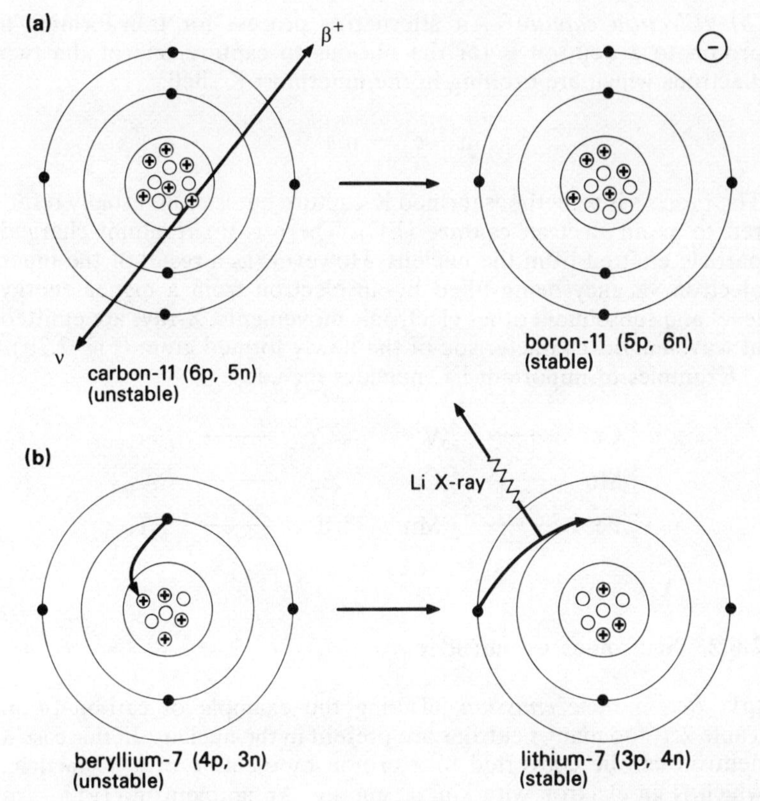

Figure 2.2 (a) Carbon-11 undergoing radioactive disintegration to give boron-11, a positron and a neutrino (the valency electrons rearrange during subsequent chemical changes). (b) Beryllium-7 undergoing electron capture to give stable lithium-7. The process is accompanied by the emission of an X-ray whose wavelength is characteristic of the lithium atom.

As a result of the loss of one positive charge in the nucleus a new nucleus is formed which occurs one place back in the periodic table. However, the total number of nucleons remains unchanged, so that the mass number *A* remains the same. Some examples of important positron emitting nuclides are

$$^{11}_{6}C \longrightarrow {}^{11}_{5}B + \beta^+$$

$$^{13}_{7}N \longrightarrow {}^{13}_{6}C + \beta^+$$

$$^{22}_{11}Na \longrightarrow {}^{22}_{10}Ne + \beta^+$$

$$^{65}_{30}Zn \longrightarrow {}^{65}_{29}Cu + \beta^+$$

(b) Electron capture An alternative process for transforming a proton to a neutron is for the nucleus to capture one of the two electrons which are orbiting in the innermost K shell:

$$p + e^- \rightarrow n + \nu$$

The process is sometimes termed K-capture but is more usually referred to as an **electron capture** (EC). There is no resultant charged particle emitted from the nucleus. However, as a result of the inner electron vacancy being filled by an electron from a higher energy level and subsequent other electronic movements, X-rays are emitted at wavelengths characteristic of the newly formed atom (Fig. 2.2b).

Examples of important EC nuclides include

$$\begin{array}{llll}
{}^{51}_{24}\mathrm{Cr} \longrightarrow {}^{51}_{23}\mathrm{V} & & {}^{57}_{27}\mathrm{Co} \longrightarrow {}^{57}_{26}\mathrm{Fe} \\
{}^{54}_{25}\mathrm{Mn} \longrightarrow {}^{54}_{24}\mathrm{Cr} & & {}^{75}_{34}\mathrm{Se} \longrightarrow {}^{75}_{33}\mathrm{As} \\
{}^{55}_{26}\mathrm{Fe} \longrightarrow {}^{55}_{25}\mathrm{Mn} & & {}^{125}_{53}\mathrm{I} \longrightarrow {}^{125}_{52}\mathrm{Te}
\end{array}$$

2.6.2 Neutron-excess nuclides

(a) Beta-particle emission Taking the example of carbon-14 in Table 2.1, too many neutrons are present in the nucleus. In this case a neutron can be converted to a proton by ejection of a **β^--particle**, which is an electron with kinetic energy. An antineutrino ($\bar{\nu}$) is also ejected:

$$n \rightarrow p + e^- + \bar{\nu} + \text{energy}$$

As the nucleus gains one positive charge the product nuclide is to be found one place to the right in the period table, although again the mass number is unchanged. Many of the biologically most interesting nuclides are β^--particle emitters:

$$\begin{array}{llll}
{}^{3}_{1}\mathrm{H} \longrightarrow {}^{3}_{2}\mathrm{He} + \beta^- & & {}^{36}_{17}\mathrm{Cl} \longrightarrow {}^{36}_{18}\mathrm{Ar} + \beta^- \\
{}^{14}_{6}\mathrm{C} \longrightarrow {}^{14}_{7}\mathrm{N} + \beta^- & & {}^{45}_{20}\mathrm{Ca} \longrightarrow {}^{45}_{21}\mathrm{Se} + \beta^- \\
{}^{24}_{11}\mathrm{Na} \longrightarrow {}^{24}_{12}\mathrm{Mg} + \beta^- & & {}^{59}_{26}\mathrm{Fe} \longrightarrow {}^{59}_{27}\mathrm{Co} + \beta^- \\
{}^{32}_{15}\mathrm{P} \longrightarrow {}^{32}_{16}\mathrm{S} + \beta^- & & {}^{131}_{63}\mathrm{I} \longrightarrow {}^{131}_{64}\mathrm{Xe} + \beta^- \\
{}^{35}_{16}\mathrm{S} \longrightarrow {}^{35}_{17}\mathrm{Cl} + \beta^- & & &
\end{array}$$

2.6.3 Heavy nuclides

(a) Alpha-particle emission Very heavy nuclides with more than 82 protons in their nuclei can attain a stable configuration by ejecting a helium nucleus of atomic mass number four (two protons and two neutrons) in the form of an **α-particle**. A typical example is the decay of radium to give radon gas:

$$^{226}_{88}\text{Ra} \longrightarrow\ ^{222}_{86}\text{Rn} + {}^{4}_{2}\text{He}^{2+} + \text{energy}$$

Some very long lived α-emitters such as ^{232}Th, ^{235}U and ^{238}U are the parents of quite long series of natural radioactive decay products. By assuming that these elements were present when the Earth was first formed it is possible to calculate from their present abundances that the age of the Earth is $\sim 4.5 \times 10^9$ a. Similarly, by measuring the relative atomic abundance ratios of these and other long lived nuclides it is possible to calculate the age of many geological and fossilized specimens. In general, however, α-particle emitters are not of much interest in biological research except in very special applications such as radiotoxicity studies or environmental investigations in the vicinity of nuclear power stations.

(b) Nuclear fission A few very heavy nuclides can undergo **fission** into two or more fragments with the release of several neutrons and large amounts of energy. The fission reaction can be either spontaneous or, more usually, induced by bombardment with nuclear particles. Thus, fission of one uranium atom can release around 5×10^6 times as much energy as can be obtained by combustion of a carbon atom in oxygen:

$$^{235}_{92}\text{U} + \text{n} \longrightarrow\ ^{236}_{92}\text{U} \longrightarrow \begin{cases} ^{90}_{38}\text{Sr} \\ 3\text{n} + \sim 200\ \text{Mev} \\ ^{143}_{54}\text{Xe} \end{cases}$$

compared with

$$\text{C} + \text{O}_2 \longrightarrow \text{CO}_2 + 4\ \text{eV}$$

The extra neutrons from the fission reaction can be used to propagate a 'chain reaction' of further fissions. This can result in a massive explosion (the atomic bomb) or, if carefully controlled (a nuclear reactor), in the release of abundant energy for peaceful purposes such as electricity generation. The isotopes of strontium and xenon in the above example are only two of at least 250 different **fission products**. In general the fission is slightly assymetric and the products fall into two groups of mass numbers 72–117 and 118–161, peaking at approximately 10% yield around mass numbers 95 and 140. Some of the most abundant fission products are ^{85}Kr, ^{90}Sr, ^{131}I and ^{137}Cs, and these are amongst the most obnoxious constituents of 'fallout' from nuclear bombs.

2.6.4 Other types of radiation

(a) Gamma-radiation The energy possessed by nuclei is quantized and, just like electrons, nuclei can exist in the ground state and in various discrete excited states. After a radioactive distintegration many nuclei are still in an excited energy state and decay to the ground state by the emission of photons of electromagnetic radiation known as **γ-rays**. This radiation is similar to X-rays but of shorter wavelength (0.001–0.1 nm) and is characteristic of the nucleus from which it arises. Thus, ^{24}Na is a β-particle emitter and decays to a highly excited ^{24}Mg nucleus which immediately decays to the ground state with the emission of two γ-rays of 1.37 and 2.75 MeV.

(b) Isomeric transition (IT) Occasionally the excited nuclei of a particular nuclide do not dissipate their energy instantly. They exist in a metastable state and decay by emitting γ-photons in a random manner over a finite period of time. Nuclides with the same A and the same Z but in different energy states are called **isomers**, the more excited isomer being denoted by the letter 'm' following the mass number in its symbol. Thus, ^{80m}Br is an isomer of ^{80}Br.

(c) Internal conversion electrons Another form of radiation arises when an excited nucleus transfers its excess energy to an orbital electron instead of emitting a γ-photon. The electron is ejected from the atom as a particle which possesses the discrete energy associated with that particular transition. **Internal conversion** (IC) can be partial or complete. An example of both instances can be found in Section 2.7, Figure 2.4 (pp. 22–3). The resultant electron vacancy has to be filled and this process gives rise to additional X-rays.

(d) Auger electrons Auger electrons arise in a similar manner to IC electrons except that the energy imparted to them originates from electronic transitions which normally give rise to X-rays. A crude analogy would be to regard them as internally converted X-rays.

2.6.5 Rate of radioactive decay

(a) The decay equation A collection of radioactive atoms will disintegrate one by one until there is none left. In some cases this is a

very rapid process while in others it can take billions of years. The events do not occur at regular intervals but in a totally random fashion which can be treated by statistical analysis.

For any given nuclide the probability that an atom will disintegrate in time t is given by the **decay constant**, λ.

If at a time t_0 there are n_0 radioactive atoms and a small number dn disintegrate after a time $t_0 + \mathrm{d}t$, then

$$\mathrm{d}n = -\lambda n_0\,\mathrm{d}t \tag{2.1}$$

Equation 2.1 is known as the decay equation, it can be rearranged to give

$$\text{rate of disintegration} = \frac{\mathrm{d}n}{\mathrm{d}t} = -\lambda n_0 \tag{2.2}$$

The rate of disintegration is known as the **activity**, a:

$$a = \frac{\mathrm{d}n}{\mathrm{d}t} \tag{2.3}$$

It is important to note that the rate of the decay process *cannot* be influenced by any external agency such as heat, light, pressure, chemical reaction or even the proximity of other radioactive atoms undergoing disintegration. So, boiling a solution of a radioactive substance may cause chemical decomposition but it will have no effect whatsoever on the radioactivity.

(b) The concept of half-life Radioactive elements decay exponentially as illustrated in Figure 2.3, where the time axis shows units of time required for a given amount of the activity to decay to exactly half its original value. The concept of half-life, originally introduced by Rutherford in 1904, is much more useful than the fundamental decay constant. The two are related, as can be shown by integrating the decay equation (2.1) in the form

$$\int_{n_0}^{n_t} \frac{\mathrm{d}n}{n} = -\lambda \int_0^t \mathrm{d}t$$

to give

$$\ln\left(\frac{n_t}{n_0}\right) = -\lambda t \tag{2.4}$$

then substituting $n_t = n_0/2$ when $t = t_{\frac{1}{2}}$ (the half-life) gives

$$\ln 2 = -\lambda t_{\frac{1}{2}}$$

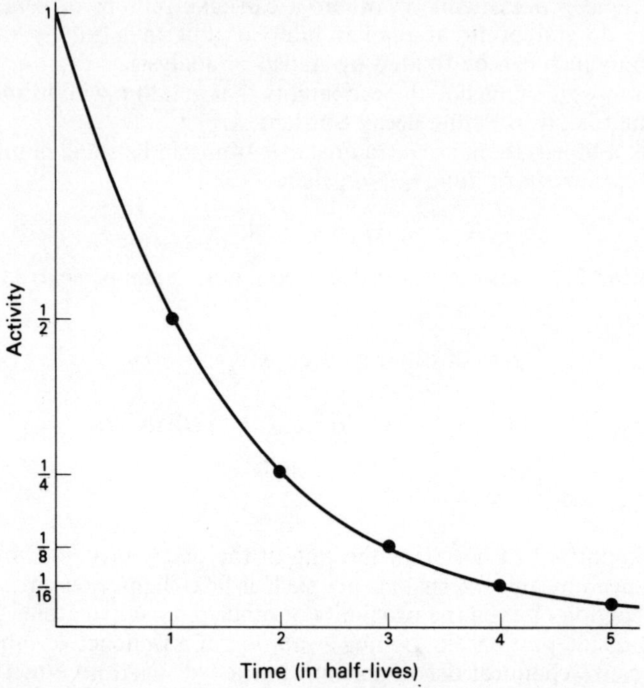

Figure 2.3 Radioactive decay in terms of half-life. The curve is exponential and shows quite clearly that whatever the activity is at the start of an experiment, short-lived radionuclides will be difficult to measure after 5 or 6 half-lives.

or

$$t_{\frac{1}{2}} = \frac{\ln 2}{\lambda} \tag{2.5}$$

Clearly, the half-life has a specific value for a specified nuclide depending on the value for λ and may be used as an aid for the identification of unknown radioactive sources. Half-life is also of considerable importance when considering the design of experiments and the disposal of radioactive waste. The half-lives of a few important isotopes are given in Table 2.2 to illustrate these points. Thus, chlorine-36 is a useful β^--emitting isotope which can be used as a tracer for chlorine. The very long half-life, however, introduces a limitation in terms of a relatively low specific activity for labelled compounds. One reason for this is the very low rate of disintegration

Table 2.2 Half-lives of some important isotopes.

Isotope	*Half-life*
^{36}Cl	303 000 a
^{14}C	5730 a
^{90}Sr	28 a
^{3}H	12.26 a
^{35}S	87 d
^{131}I	8.04 d
^{24}Na	15 h
^{13}N	10 min
^{15}O	2 min

per unit weight of nuclide, but a less obvious reason is the need for long and expensive reactor irradiation times required to produce the nuclide. The same is true, although to a lesser extent, for ^{14}C-labelled compounds which cannot be obtained with such high specific activity as, for example, tritium compounds. Thus, where the highest sensitivity is required, tritium is often used in preference to carbon-14, but there are hidden dangers in this practice which are discussed in Chapter 5.

The time required for experiments becomes a problem when nuclides have half-lives measured in terms of a few days or hours. Generally speaking an isotope is only useful for periods extending to around five half-lives. In biological science, one of the tragedies of nature is that nitrogen-13 and oxygen-15 are the longest lived radioactive isotopes of these elements and that these isotopes have half-lives which are far too short for most practical purposes in biological research.

The isotopes strontium-90 and iodine-131 are representative of the various nuclides which occur in 'radioactive fallout' from nuclear weapons. The iodine presents a relatively short term, high activity hazard, whereas strontium, which can replace calcium in bone, is a more insidious long term hazard extending to future generations.

(c) Activity and specific activity Activity was defined by equation (2.3) and there are two units of activity. The **curie** (Ci) is defined as 3.7×10^{10} disintegrations per second. This is the only unit to appear in the scientific literature up to 1977/78. It represents rather a large amount of activity and biological experiments rarely require more than a few micro- or millicuries. The newer SI unit of activity goes to the other extreme. The **becquerel** (Bq) is defined as 1 disintegration

per second. Thus, it is now necessary to report activities in kilo- or megabequerels.

Specific activity is most correctly defined as the ratio of the number of radioactive atoms to the total number of atoms of that element in the given compound. In practice the term specific activity is used much more loosely to represent activity per unit weight, which can be expressed in many ways such as millicuries per mole, microcuries per gram, kilobequerels per micromole, etc.

2.7 Decay schemes

Much of the information discussed so far in this chapter can be represented by simple diagrams in which vertical distances represent changes in nuclear energy levels (Q) and moves to the right or left represent gain or loss of positive nuclear charge (Z). These plots of energy against atomic number can also incorporate additional information on half-lives, energies of different particles and relative importance (as a percentage) of different decay pathways. Some typical examples of various simple and one complex decay schemes are given in Figure 2.4.

2.8 Properties of radiations

The particles from radioactive disintegration are frequently referred to as **ionizing radiations** since their major effect on matter is to 'knock' electrons out of orbit to form ion pairs. A related but less prominent effect is **excitation**, where electrons in orbit are excited to higher energy states and then emit ultraviolet or visible light as they return to the ground state. Both effects are extremely important as they form the basis of all the methods of detection. Excitation is also the source of light in the luminous paint used on clock, watch and instrument dials.

In air, about 34 eV are required to produce one ion pair. Thus a particle loses this amount of energy at each ionizing event and will eventually lose all its energy and come to rest. The amount of ionization caused by a particle obviously depends upon the initial energy, but the distance travelled depends upon a second factor – the **specific**

Figure 2.4 Some typical decay schemes which illustrate most of the processes described in the text. The scheme for metastable bromine-80 is quite complex but by no means untypical of many man-made isotopes. EC = electron capture; IC = internal conversion; IT = isomeric transition. The vertical lines associated with position (β^+) emission represent energy equivalent to the two electron masses (2×0.51 MeV) which are lost in this decay process.

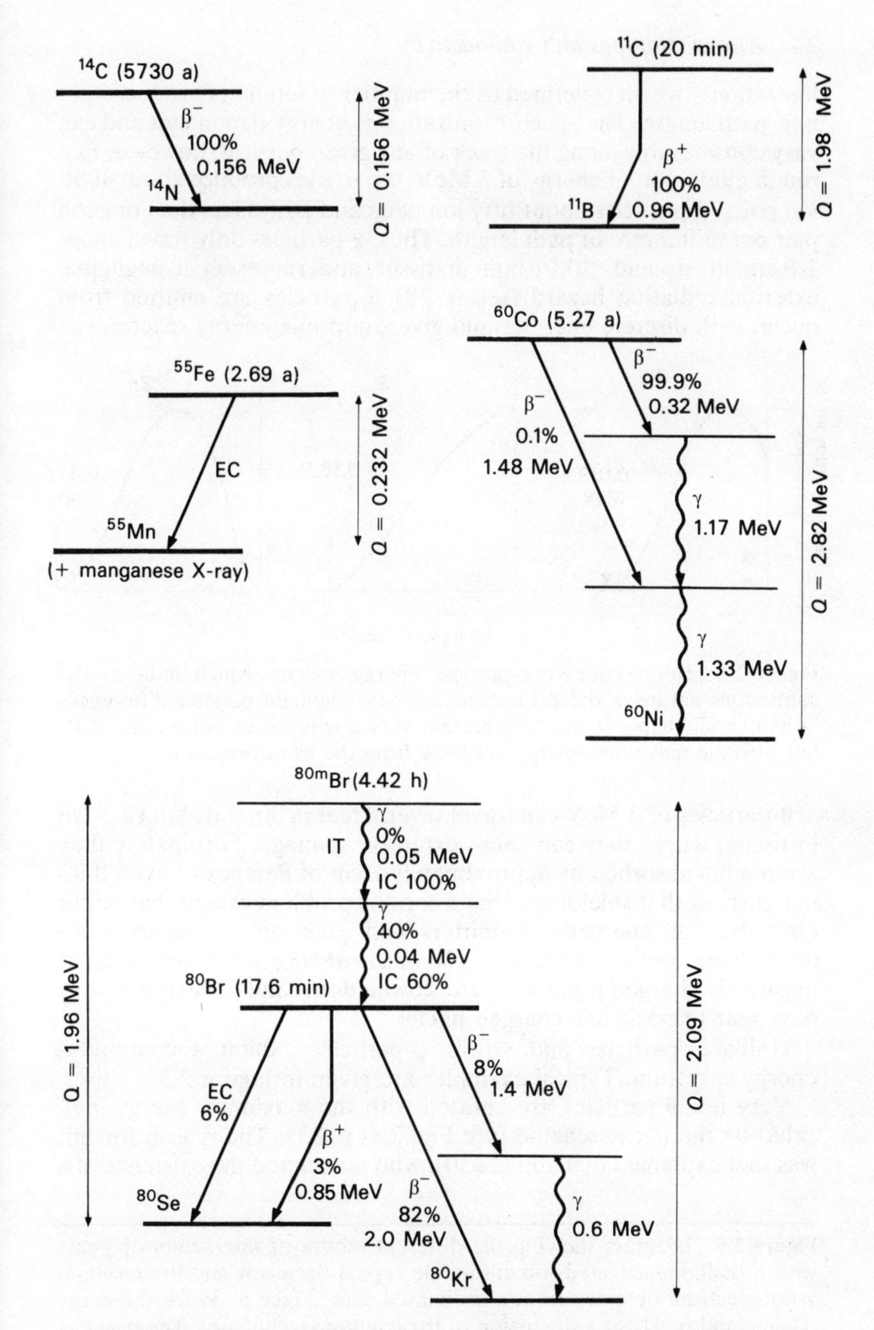

^{14}C (5730 a)
β⁻
100%
0.156 MeV
^{14}N
Q = 0.156 MeV
^{11}C (20 min)
β⁺
100%
^{11}B
0.96 MeV
Q = 1.98 MeV
^{55}Fe (2.69 a)
EC
^{55}Mn
Q = 0.232 MeV
(+ manganese X-ray)
^{60}Co (5.27 a)
β⁻
99.9%
0.32 MeV
β⁻
0.1%
1.48 MeV
γ
1.17 MeV
γ
1.33 MeV
^{60}Ni
Q = 2.82 MeV
^{80m}Br (4.42 h)
γ
IT
0%
0.05 MeV
IC 100%
γ
40%
0.04 MeV
IC 60%
^{80}Br (17.6 min)
Q = 1.96 MeV
EC
6%
β⁺
3%
0.85 MeV
^{80}Se
β⁻
82%
2.0 MeV
β⁻
8%
1.4 MeV
γ
0.6 MeV
^{80}Kr
Q = 2.09 MeV
(+ bromine X-rays, + selenium X-rays, + 0.51 γ annihilation radiation)

ionization – which is defined as the number of ion pairs produced per unit path length. The specific ionization is energy dependent and can vary considerably along the track of any given particle; however, as a rough guide, for an energy of 3 MeV, α-particles produce about 4000 ion pairs, β-particles about fifty ion pairs and γ-rays less than one ion pair per millimetre of path length. Thus, α-particles only travel about 1.6 cm in air and <0.01 mm in tissue and represent a negligible external radiation hazard (see p. 38). α-particles are emitted from nuclei with discrete energies and give sharp line energy spectra.

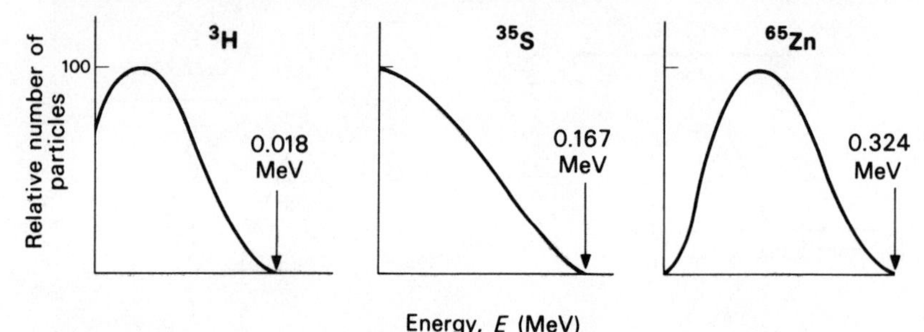

Figure 2.5 Representative β-particle energy spectra which indicate the continuous nature of the β-spectrum and also highlight possible differences in its overall shape. It can be seen that very few particles indeed are emitted with the maximum energy available from the transformation.

β-particles of 3 MeV can travel several feet in air and about 1.5 cm in tissue, where they can cause extensive damage. Fortunately they are readily absorbed by approximately 1 cm of Perspex or even thinner glass so that shielding is not a serious problem except that larger quantities of energetic β-emitters can give off secondary electromagnetic radiation known as ***brehmsstrahlung*** which arises when negatively charged β-particles are slowed down and deflected as they pass near to positively charged nuclei.

Unlike α-particles and γ-rays, β-particles exhibit a continuous energy spectrum. Typical examples are given in Figure 2.5.

Very few β-particles are emitted with the maximum energy indicated by the decay scheme (see Fig. 2.4; p. 23). The reason for this was first explained by Pauli (1930), who postulated the existence of a

Figure 2.6 Diagram showing the different modes of interaction of γ-rays with a thallium-activated sodium iodide crystal dectector and the resultant γ-ray spectrum obtained from a sodium-24 source (see p. 39 for the decay scheme and p. 51 for a discussion of the counting technique). The peaks at 1.38 and 2.76 MeV are due to deposition of all the energy from the

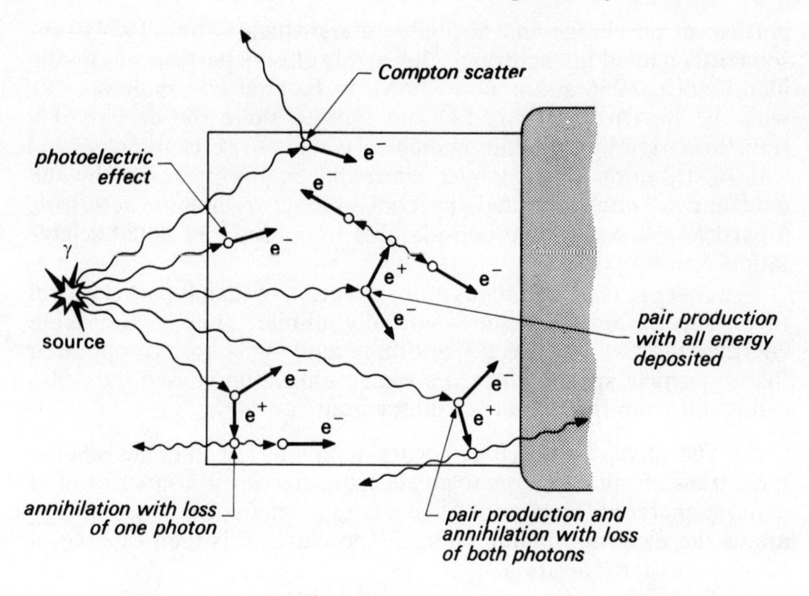

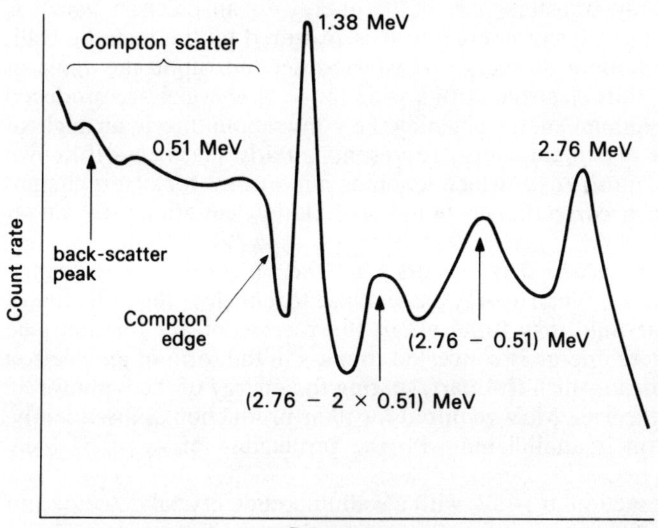

γ-photons. The peaks at 2.25 and 1.74 MeV are due to pair production followed by loss of one or two of the positron annihilation photons as shown. The peak at 0.51 MeV is due to annihilation radiation. The lowest energy peak is due to γ-rays which have been back-scattered from the surrounding shield and is often indistinguishable from the general Compton scatter.

particle of no charge and negligible mass which Fermi (1934) subsequently named the neutrino. This highly elusive particle was finally identified in 1956 and is now known to be emitted simultaneously with the β-particle (see p. 14) and thus to share the energy of a transformation in a random manner. This gives rise to the observed wide distribution of β-energies since the energy possessed by the neutrino is not deposited in conventional radiation detectors. β-particle spectra assume considerable importance in liquid scintillation counting (p. 57).

γ-rays are at least 100 times more penetrative than β-particles, and their range in air and tissue is virtually infinite. They have discrete energies and give line spectra, but these tend to be more complicated than α-particle spectra since the sparse ionization caused by γ-rays can result from three kinds of interaction:

(a) The **photoelectric effect** occurs mainly at low energies when a γ-ray transfers all its energy to an electron, ejecting it from its orbit as a monoenergetic electron whose energy equals the γ-ray energy minus the electron binding energy. The electron is then capable of causing further ionization.

(b) **Compton scattering** occurs in the intermediate energy range when a γ-ray transfers part of its energy to an electron which is ejected while a lower energy γ-ray is scattered by the inelastic collision. The amount of energy transferred depends upon the angle of scattering, thus electrons with a wide range of energies are produced up to a maximum energy given by the γ-ray rebounding at an angle of 180°. This maximum energy represents a fairly sharp cut-off known as the Compton edge, which assumes importance for external standard quench correction in liquid scintillation counting (see Ch. 4, p. 60).

(c) **Pair production** occurs in the higher energy range (>1.02 MeV). When a γ-ray passes close to a nucleus the influence of the nuclear field may bring about the reverse of annihilation (see p. 15), where energy is converted to mass in the form of an electron and a positron which fly apart, sharing the energy of the γ-photon in excess of the 1.02 MeV required for their production. Subsequently, the positron is annihilated with the production of two 0.51 MeV γ-rays.

The interaction of γ-rays with a sodium iodide crystal detector and the derived γ-ray spectrum for sodium-24 are illustrated in Figure 2.6. In all three processes the original γ-ray disappears so that γ-rays cannot be said to give tracks like the α- and β-particles. Attenuation of γ-rays to reduce radiation hazards requires a considerable thickness of dense material such as lead or barium concrete.

2.9 Practical experiments

2.9.1 Penetrating power of radiations and the energy of β-particles from a ^{32}P source

(a) Apparatus and materials

(1) Thin mica end window-type Geiger–Müller (GM) counter mounted in a lead castle with source mounting shelves beneath (see Fig. 2.7)
(2) High voltage, scaler, timer and quenching probe unit appropriate for the GM tube used
(3) Standard source of carbon-14 (disc of labelled poly(methylmethacrylate))
(4) A set of calibrated aluminium foil absorbers
(5) An aqueous solution of phosphorus-32 containing approximately 925 kBq cm^{-3} (25 μCi cm^{-3}).

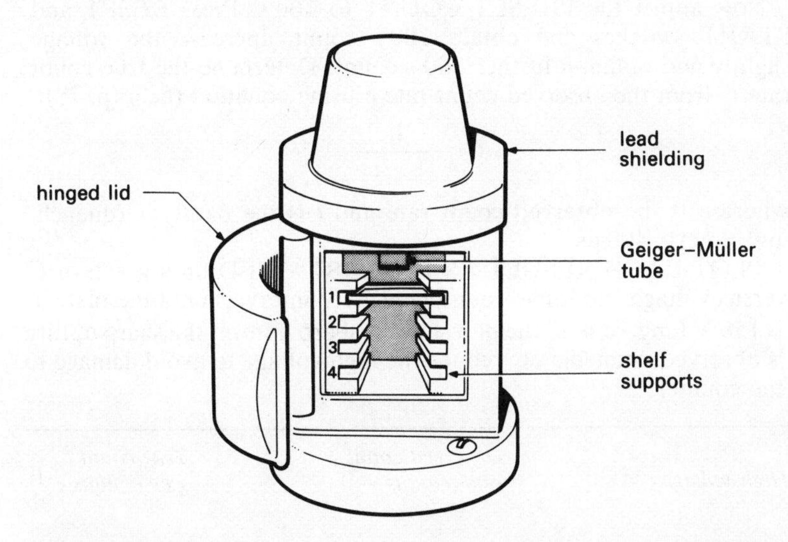

Figure 2.7 Thin mica end window-type Geiger–Müller counter mounted in a lead castle with source mounting shelves beneath. The surrounding lead acts as a shield against external radiation sources. The radioactive source can be placed at different distances from the counter by selecting one of four different positions.

(b) Setting the high voltage on the GM tube This must be set to the appropriate level either by following the method described below or by reference to the instrument instructions.

If you are using the following method, the controls on the instrument must first be adjusted before switching on the power. The way in which these controls are set will depend upon the machine available, but the following list should be adhered to as closely as possible:

Display	Count
Preset count	Off
Preset time	Off
Threshold	Minimum
High voltage	0–1 kV, zero
GM quenching time	400 μs

Now switch on the power, place the ^{14}C source on shelf 3, press RESET switch and check that all tubes show zero.

Press START switch, then slowly increase the high voltage until counts first appear on the scaler – this is the STARTING VOLTAGE.

Now adjust the PRESET COUNT to 100 s. Press RESET and START switches and obtain 100 s count. Increase the voltage slightly and obtain a further 100 s count. Determine the true count rate C from the observed count rate c using equation (5.1) (p. 78):

$$C = \frac{c}{1 - ct}$$

where c is the observed count rate and t is the paralysis (quenching) time = 400 μs.

PLOT EACH RESULT AS IT IS OBTAINED on a graph of C versus voltage. Continue counting at 30 V intervals until the plateau is 150 V long, or until the plot shows a sharp upturn. If a sharp upturn is observed, immediately reduce the high voltage to avoid damage to the counter.

High voltage	*Observed count rate (counts s^{-1})*	*True count rate (counts s^{-1})*
.	.	.
.	.	.
.	.	.

From your plot determine the starting voltage, threshold voltage, length of plateau and slope of plateau (see Ch. 4, p. 47).

(c) Measurement of the maximum energy of β-particles from a ^{32}P source The absorption of β-particles by matter is almost independent of the material, provided the thickness is expressed in units which are a function of density, e.g. milligrams per square centimetre. Thus, 0.5 mm of lead absorbs β-radiation to the same extent as 2 mm of aluminium – each has a thickness of 570 mg cm^{-2}.

Prepare two ^{32}P sources, one 50 times stronger than the other. First, pipette 0.2 cm^3 of the ^{32}P stock solution into an aluminium planchet (use a safety pipette for this procedure). Then, using the same pipette, transfer 0.2 cm^3 to a 10 cm^3 volumetric flask and make up to the mark with distilled water containing a little inactive phosphate 'carrier'. Use a second clean pipette to transfer 0.2 cm^3 of this solution to a planchet. Carefully dry both planchets under an infrared lamp.

Place the low activity source on shelf 2 in the lead castle and count for a pre-set 10 000 counts. Place an aluminium absorber of about 20 mg cm^{-2} on shelf 1 and count again. Continue counting at increasing absorber thickness, switching to a pre-set count of 3000 counts when expedient, until the time for 3000 counts exceeds 1000 s. Then change to the high activity source (DO NOT ALLOW THE STRONG SOURCE TO BE EXPOSED TO THE COUNTER WINDOW – ALWAYS *COVER* IT BEFORE CHANGING ABSORBER) and repeat measurements with the two previous absorbers. Calculate the average ratio between strong and weak sources. (N.B. Remember paralysis time corrections and background corrections.) This calculated factor is used to transform weak source net count rates into strong source net count rates. Continue counting with increasing thickness until an almost constant count rate, due to *brehmsstrahlung*, is obtained. Tabulate your results:

Absorber thickness	*Count rate*	*Corrected count rate, c_t (counts s^{-1})*	Log c_t
.	.	.	.
.	.	.	.
.	.	.	.

Plot the log c_t against absorber thickness. Allow for air absorption (30 mm path ~3.9 mg cm^{-2}) and counter window thickness (~1.2 mg cm^{-2}).

After plotting your results, establish a value for *brehmsstrahlung* count rate by extrapolation from count rate values at high absorber thickness (almost constant) to zero absorber thickness. Then deduct this from the values of c_t previously tabulated and re-plot the data on the same graph. The corrected curve should asymptotically approach a vertical line through the 'range point' as β-transmission becomes zero (i.e. long c_t (corrected) $\rightarrow -\infty$). The range point corresponds to the maximum β-energy of ^{32}P of 1.71 MeV ($\equiv$760 mg cm^{-2}).

(d) Measurement of the thickness of a piece of aluminium foil using β-particle absorption This simply makes use of the curves already obtained for β-particle absorption. Place a piece of household foil on shelf 1, supported if necessary by card, and use one of the sources to obtain a count rate. Read off the absorber thickness in milligrams per square centimetre from the *x*-axis. Thickness may be calculated from $F = DS$, where F is the thickness in milligrams per square centimetre, D is the density in milligrams per cubic centimetre (= 2702 mg cm^{-3}), and S is the thickness in centimetres.

2.9.2 Separation of barium-137m from its parent caesium-137 and determination of the half-life

Metastable barium-137 has a very short half-life and emits a γ-ray of 0.66 MeV. Because of the short half-life, a rapid means of separation from its long lived parent is desirable. Ion exchange on a cation exchange resin has been used successfully for separation. An alternative method, however, makes use of an anion exchange resin – Amberlite IRA-400 – on which a complex is formed with cobalt ferrocyanide; the barium may then be eluted with distilled water. The 0.66 Mev γ-ray may be measured on a crystal scintillation counter.

(a) Apparatus and materials

(1) A chromatography column (e.g. 25 cm^3 burette)
(2) Amberlite resin IRA-400
(3) Potassium ferrocyanide solution (2 mol dm^{-3})
(4) Cobalt nitrate solution (2 mol dm^{-3})
(5) Caesium-137 chloride solution (37 kBq cm^{-3})
(6) Crystal scintillation counter
(7) Stop-watch.

(b) Counting Switch on the scintillation counter set to the optimum conditions for the 0.66 MeV γ-ray. Place a counting vessel containing 10 cm^3 of distilled water on the crystal and commence a background count determination.

(c) Separation While the background is being counted proceed as follows:

(1) Loosely pack a small wad of glass wool in the bottom of the glass column. Close the tap and fill the column with distilled water until it is three-quarters full. Then add resin IRA-400 very slowly, with gentle tapping, from a folded piece of paper, until a depth of approximately 8 cm is reached. Open the tap and allow the water to drain into a beaker until the meniscus is just above the resin. NEVER ALLOW THE COLUMN TO RUN DRY.
(2) Add 10 cm^3 of potassium ferrocyanide solution to the column and drain very slowly.
(3) Wash the column with distilled water (50 cm^3) and drain to remove excess ferrocyanide.
(4) Add 10 cm^3 of cobalt nitrate solution and drain very slowly.
(5) Wash the column with distilled water (100 cm^3) and drain as before.
(6) Very carefully transfer 1.0 cm^3 of the caesium-137 solution to the resin on top of the column, drain very slowly and leave for 5 min.
(7) Wash the column with 200 cm^3 of distilled water. This removes excess caesium-137 and all the barium-137 m.
(8) Allow 30 min for barium-137m to build up again. During this period record the background. Then add 7 cm^3 distilled water to the column and (THE NEXT OPERATIONS MUST BE PERFORMED VERY RAPIDLY) drain quickly into a polythene counting vessel marked 'sample'.
(9) IMMEDIATELY transfer the pot to the crystal. Set the timer to 10 s, then simultaneously start the stop-watch and counter. Record the count for 10 s, then re-zero and at exactly 30 s on the stop-watch start the counter again. Continue in this way, recording 10 s counts every 30 s until the activity falls by a factor of 16–32.

Correct all count rates for background then plot log count rate against time and obtain the half-life.

3 Radiation protection and safe handling of radioactive materials

3.1 Nature of the hazards

When ionizing radiations pass through living tissue they can cause extensive damage. The hazard is particularly insidious since man has no sense capable of detecting radiation even at very high intensities. Madame Curie's husband was probably the first person who knowingly received a 'radiation burn', and the early experimenters with X-rays soon realized the harmful nature of such ionizing radiations. It has been estimated that by 1922 around 100 X-radiologists had died as a direct result of their work and it was around that time that the first moves to specify safety recommendations were made. Much other evidence concerning the harmful nature of radioactive materials has since been collected. For example: from girls working with luminous paint who licked their brushes and subsequently developed mouth and bone cancer; from miners digging radioactive ores and working in high levels of radon gas who subsequently developed lung cancers; from patients given large doses of X-rays for therapeutic purposes who subsequently developed leukaemia and other forms of cancer; and of course from the survivors of the two atomic bombs which ended World War II.

Somatic effects (i.e. damage to the exposed individual) are well documented and range from minor burns, dermatitis, ulcers, diarrhoea, hair loss and blood changes to most forms of cancer and eventual death. Some somatic effects such as carcinogenesis (production of cancer) are known as **stochastic effects** since the probability of the effect occurring is regarded as a function of dose without threshold (i.e. there is no evidence at present of a 'minimum dose'). **Non-stochastic effects** are those such as cataract of the eye lens, non-malignant damage to the skin, blood deficiencies, etc., for which the severity of the effect varies with dose and for which there is a threshold below which no detrimental effects are seen.

Hereditary effects are those which become manifest in the descendants of the exposed person. Studies with irradiated insects (*Drosophila*) and mammals (mice and rats) have shown that radiation-induced gene and chromosome mutations can cause a wide range of malformations of organs and tissues in the offspring. In man, the available evidence is very limited but, in general, studies on the children of groups of people who have been subjected to radiation have revealed no obvious genetic effects. Nevertheless, the results with animals suggest that at present it would be wise to view hereditary effects as stochastic and to accept that there is no threshold dose.

3.2 The development of radiation protection

Once the hazards had been clearly recognized, radiation workers organized themselves to formulate safeguards and have consistently led the field with regard to the safety of individuals and populations. The International Commission on Radiological Protection (ICRP) was formed in 1928 and was making specific recommendations long before national governments became involved in radiation health and safety legislation. Health physicists and radiation protection officers are respected members of the scientific community and their vigilance has ensured that the 'nuclear and isotope industry' and medical and biological uses of radiation and isotopes are amongst the safest fields of endeavour in which to be employed.

The current philosophy of the ICRP is set out in their Publication 26 (1977). They state that: 'The aim of radiation protection should be to prevent detrimental non-stochastic effects and to limit the probability of stochastic effects to levels deemed to be acceptable.' To this end various units and concepts have been defined and specific recommendations made.

3.3 The concept of radiation dose

In pharmacology the word 'dose' is used to indicate the amount of a substance per unit body weight which will produce a certain biological effect. As radiation produces biological effects, the term 'dose' has been adopted by analogy, but there the similarity ends, since radiation dose is defined in purely physical terms as the amount of energy absorbed per unit weight of tissue. The SI unit of **absorbed dose** is the **gray**:

$$1\,\text{Gy} = 1\,\text{J kg}^{-1}$$

The unit replaces the older unit the **rad** (1 rad = 100 erg g^{-1}). Since many instruments calibrated in rads will continue to be used for a long time, the relationship between the two units should be noted:

$$1\ \text{rad} = 10\ \text{mGy}$$

In practice it has been found that the absorbed dose alone is insufficient to predict either the severity or the probability of the deleterious effects. Some radiations are more damaging than others and hence ICRP recommends introduction of a quality factor, Q, as set out in Table 3.1.

Table 3.1 Values of quality factor used in defining dose equivalents.

Particles	Q
X-rays, γ-rays and electrons	1
neutrons, protons, etc., of unknown energy	10
α-particles and other multiply-charged particles of unknown energy	20

Q is used to calculate the dose equivalent:

$$\text{dose equivalent} = \text{absorbed dose} \times Q$$

Under the old system of units, the roentgen equivalent man or **rem** was defined as the number of rads $\times Q$. The SI unit of dose equivalent is the **sievert** (Sv), where

$$1 \text{ rem} = 10 \text{ mSv (or } 1 \text{ Sv} = 100 \text{ rem)}$$

When attempting to assess a radiation hazard it is essential to know the rate at which radiation is being absorbed, and hence we have the concept of dose rate, which is usually expressed as dose per hour, i.e. microsieverts per hour, micrograys per hour, rads per hour, etc.

3.4 Recommended dose equivalent limits

To prevent non-stochastic effects and to limit stochastic effects to acceptable levels, the ICRP recommends, for radiation workers, an annual dose equivalent limit for uniform irradiation of the whole body of 50 mSv (5 rem), although individual parts of the body may receive higher doses. The recommended dose limit for members of the general public is 5 mSv (0.5 rem) per year. Bearing in mind the very limited opportunities for individual exposure the average dose to the population as a whole will be extremely small. Despite these upper limits, there is an over-riding philosophy that doses should be kept as low as is readily achievable, and in practice very few radiation workers ever receive anything like the permissible dose.

3.5 Secondary limits and derived limits

Sources of radiation external to the body can readily be monitored with suitable instruments and the potential dose equivalent assessed.

However, when open sources of radioisotopes are used there is the possibility of ingestion into the body. The subsequent assessment of dose is greatly complicated by the uneven distribution and metabolism of the material. To meet these difficulties the ICRP publishes a list of **annual limits of intake** (ALI) by inhalation or ingestion. These are secondary limits designed to ensure that the internal dose does not exceed the recommended dose equivalent limit. Some examples are given in the list of nuclides in Appendix B.

Finally, by taking account of the water consumed, or the air breathed, in a 'working year' it is possible to calculate derived limits such as the maximum permissible concentrations in air or in drinking water. By making certain other assumptions regarding the physical transfer of activity from, for example, bench surfaces into the air or on to the hands and hence into the mouth, it is possible to calculate maximum permissible levels of surface contamination in a radioisotope laboratory such that the dose equivalent limit cannot be exceeded. The current British National Radiological Protection Board (NRPB) recommendations are set out in Table 3.2 overleaf.

3.6 Radiation protection and student experiments

ICRP Publication 13 (1968) deals specifically with radiation protection in schools for pupils up to the age of 18 years. The recommendations are entirely suitable for adoption for undergraduate class experiments in universities. For tracer experiments it is recommended that the amount of activity used per experiment should never exceed the limit for ingestion or inhalation, whichever is the less, by a member of the public in a year. Some typical examples are given in Appendix B. Almost all suitable experiments can be carried out with activities much smaller than those listed, and if activities are kept to the practical minimum then the hazards are virtually negligible. Nevertheless, good working practice will further ensure safety and this is dealt with in the final section of this chapter.

3.7 Radiation monitors

For assessing the radiation emanating from a source or for measuring surface contamination, portable versions of the ionization chambers, Geiger–Müller counters or scintillation counters described in Chapter 4 are used. For weak β-particle emitters such as ^{14}C or ^{35}S a Geiger–Müller counter with a very thin mica window is required. Surface contamination by tritium can only be detected by wiping the surface with tissue paper moistened with solvent. The 'wipe' is then

Table 3.2 A proposed classification of radionuclides for controlling surface contamination in workplaces.

Category	Surface	Extent of contamination	Class I	Class II	Class III	Class IV	Class V
		Levels of contamination that should not be exceeded ($Bq\ cm^{-2}$)					
A	Surfaces of the interiors and contents of glove boxes and fume cupboards	← the minimum reasonably achievable →					
B	Surfaces of active areas and of plant, apparatus, equipment (including personal protective clothing), materials and articles within active areas other than those in category A	$<1\ m^2$	3	3×10^1	3×10^1	3×10^2	3×10^3
		$>1\ m_2$	3×10^{-1}	3			
C	Surfaces of the body		3×10^{-1}	3×10^{-1}	α-Emitters: 3×10^{-1}; Others: 3	3×10^1	3×10^2
D	Inactive areas, personal clothing		3×10^{-1}	3×10^{-1}	3	3×10^1	3×10^2

Notes

1. In relation to category C (surfaces of the body) the values for all radionuclides may be increased by a factor of 10 when the skin is monitored with a small area probe.
2. For ^{231}Pa use a tenth of the values in class I for surfaces in categories B, C and D. For ^{237}Np use a tenth of the values in class I for surfaces in category C.
3. These values are not applicable to volatile compounds and radionuclides in forms that can readily penetrate the skin.
4. Direct monitoring should be employed wherever practicable. If wipe testing is employed, the assumption should be made that 10% of the contamination has been removed.
5. Radionuclide classes are as follows:

I: ^{227}Ac, ^{228}Th, ^{230}Th, ^{232}Th, Th-nat, ^{231}Pa, ^{232}U, ^{233}U, ^{234}U, ^{236}U, α-emitters with $Z > 92$

II: ^{147}Sm, ^{210}Pb, ^{227}Th, ^{235}U, ^{238}U, U-depl, U-nat, U-enr, ^{241}Pu

III: Other nuclides except those in classes IV and V

IV: ^{14}C, ^{35}S, ^{54}Mn, ^{57}Co, ^{65}Zn, ^{67}Ga, ^{75}Se, ^{77}Br, ^{85}Sr, ^{99m}Tc, ^{109}Cd, ^{123}I, ^{125}I, ^{129}Cs, ^{197}Hg

V: ^{3}H, ^{51}Cr, ^{55}Fe, ^{63}Ni, ^{131}Cs

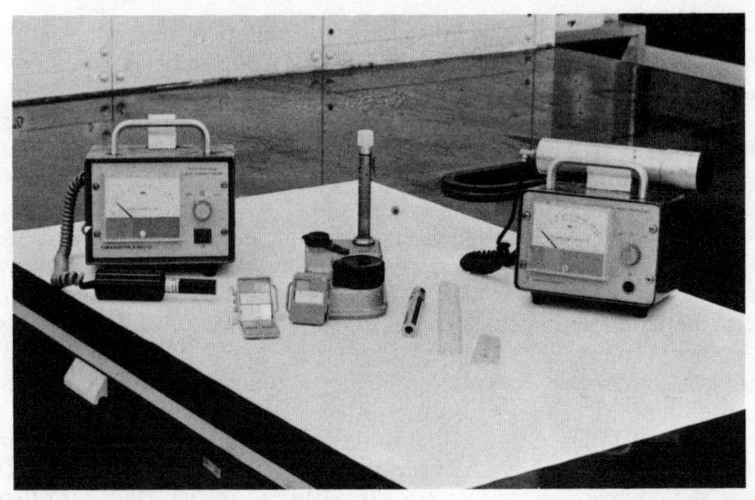

Figure 3.1 Various radiation monitoring devices. The instrument on the left has a small Geiger–Müller probe and is used for measuring dose rates. The instrument on the right has a thin sodium iodide crystal and photomultiplier probe and is suitable for detecting contamination by ^{125}I and similar electron capture nuclides. In the centre are two pocket ionization chambers, one of which is mounted on its re-charging device. Finally, at the front are two film badge holders suitable for pinning to laboratory coats and two thermoluminescent dosemeters in the form of finger stalls for measuring doses to the hands.

placed in a vial, immersed in scintillation solution and assayed in a liquid scintillation counter.

For personnel monitoring, pocket ionization chambers, film badges or thermoluminescent dosemeters are used. Such devices are only required where relatively large amounts of high energy β-emitters (e.g. ^{32}P) or γ-rays are being used. They are unable to detect weak β-particles and consequently are not necessary when ^{3}H, ^{14}C or ^{35}S are being handled. A variety of monitoring devices is shown in Figure 3.1.

3.8 Procedures for handling open sources of radioisotopes

Work with radioisotopes in any given institution should be covered by local rules which ought to take due account of the recommendations of the ICRP and also of the national legislation of the country in which the institution is situated. Most workplaces will have a radiation protection advisor, health physicist or safety officer, who should be consulted at an early stage when work with radioisotopes is contemplated.

Three separate hazards can be identified: (a) the external radiation health hazard (b) the internal radiation health hazard (ingestion), and (c) the hazard to experimental results due to a spread of contamination. The latter is not a health hazard, but when workers have been apprised of the danger to their experimental results it is often surprising how their technique and general cleanliness improves, with an obvious reduction in the likelihood of ingestion. The ingestion hazard also depends upon the nuclide. The International Atomic Energy Agency (IAEA) has defined the four groups of high, medium (upper), medium (lower) and low toxicity nuclides. Thus, plutonium-239 is a high toxicity nuclide with a specified annual limit of intake (ALI) of 200 kBq (5.40 μCi). Carbon-14 is of medium (lower) toxicity with an ALI of 200 MBq (5.4 mCi) and tritium has low toxicity with an ALI of 3×10^3 MBq (81 mCi!). Data for other useful nuclides are included in Appendix B.

3.8.1 The external radiation hazard

Since this book is concerned largely with the use of isotopes as tracers, the external radiation hazard is most likely to be a problem when dealing with new deliveries of radiochemicals. Even then weak β-emitters such as ^{3}H, ^{14}C and ^{35}S pose negligible danger since their radiations are absorbed by their containers, or, if exposed as open sources, by a few centimetres of air or the outer layers of the skin.

Hard β-emitters or γ-emitters, however, are more dangerous and their containers should be shielded and handled with simple remote control devices such as forceps or tongs. If it is assumed that a source is held in tongs of length 30 cm and that at arm's length it will be approximately 1 m from the body, then the following equations will give a very rough indication (generally erring on the safe side) of the dose rates involved:

$$\beta\text{-dose rate at } 30\text{ cm} = 80 \times A\ \mu\text{Sv h}^{-1} \quad (3.1)$$

$$\gamma\text{-dose rate at } 30\text{ cm} = 1.6 \times A \times E\ \mu\text{Sv h}^{-1} \quad (3.2)$$

$$\gamma\text{-dose rate at } 100\text{ cm} = 0.15 \times A \times E\ \mu\text{Sv h}^{-1} \quad (3.3)$$

where A is the activity in megabequerels (MBq) and E is the total γ-ray energy in megaelectronvolts (MeV) per nuclear disintegration. These equations are useful when initially discussing the design of an experiment, but they are no substitute for measurements using a properly calibrated dose rate meter when a newly purchased source is initially opened. The hazard is then controlled using a combination of **distance**, **time** and **shielding**.

A good example would be an experiment involving an aqueous solution of sodium-24:

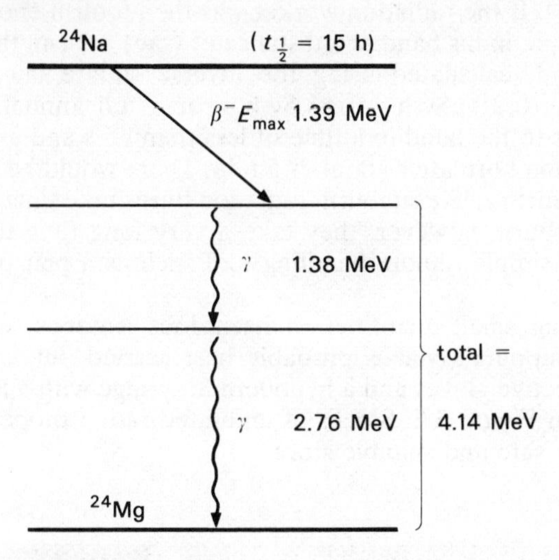

With such a short half-life it is not unusual to purchase 370 MBq (10 mCi) for a single experiment. The material would be delivered in a shielded container but when the sample vial is removed from this shield, using tongs, the γ-dose rate at 1 m would be approximately 170 μSv h^{-1} (eq. 3.3 gives $0.15 \times 370 \times 4.14 = 230$ μSv h^{-1}). Taking the annual dose equivalent limit of 50 mSv, which may reasonably be subdivided into 1 mSv per week, it is readily seen that, even taking the higher figure of 230 μSv h^{-1}, a radiation worker would be permitted to handle the source for approximately 4 h ($1 \times 10^3 \div 230$) in any one week. If, however, it was necessary to move closer, say to 30 cm, the dose rate would increase to about 2000 μSv h^{-1} (eq. 3.2 gives 2450 μSv h^{-1}) and his time would be limited to 30 min unless some form of shielding was employed.

If the calculations show that an experiment will give rise to excessive dose rates, then shielding must be used and probably more intricate procedures will be necessary which are not discussed in this book (see Appendix D).

The distance factor is crucial, ***dose varies with the inverse square of the distance***. This is dramatically demonstrated for β-emitters. In the ^{24}Na example above, the β-dose rate was not considered since most of the β-particles would be absorbed in the aqueous solution and the walls of the vial. If, however, a tenth of the sample (37 MBq, 1 mCi) was removed from the vial and evaporated to dryness in an open dish, the β-dose rate at 30 cm would be 2600 μSv h^{-1} (eq. 3.1 gives 3000 μSv h^{-1}), to which must be added the γ-dose rate of 200 μSv h^{-1}. If the radiation worker was then foolish enough to pick up the sample in his hand at a distance of (say) 0.2 cm, the dose rate to the hand, calculated using the inverse square law, would be $(2800 \times 30^2/0.2^2)$ μSv h^{-1} = 63 Sv h^{-1}, or a full annual dose equivalent limit to the hand in a time of less than 45 s and a nasty localized radiation burn after about 2–5 min. There would be no immediate discomfort as, like sunburn, radiation burns take time to develop. Unlike sunburn, however, they take a very long time to heal. The need for a simple remote handling tool such as a pair of forceps is self-evident.

Dispensing small quantities of hazardous isotopes such as ^{24}Na from the supplier's vial is probably best carried out fairly rapidly using protective gloves and a hypodermic syringe with a long needle, as shown in Figure 3.2. Stocks of undiluted radioisotopes should be locked in a safe and suitable store.

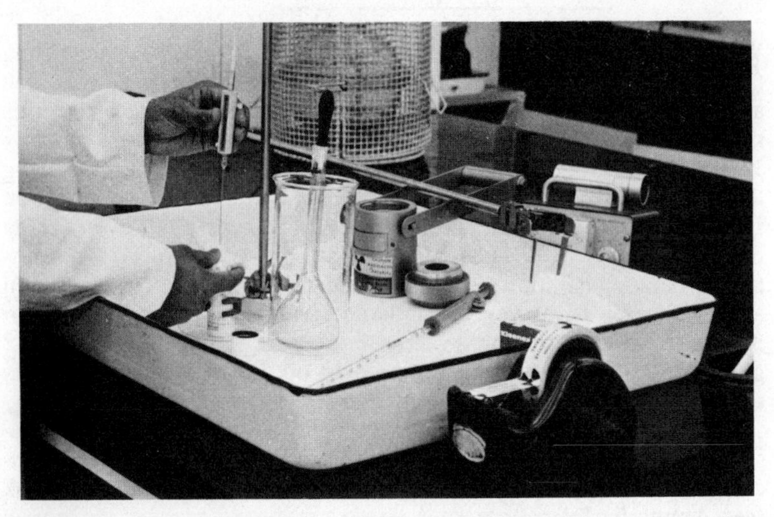

Figure 3.2 Dispensing a small quantity of a hard β-particle or γ-ray emitter from the supplier's multidose vial. Note in particular: the firmly clamped lead pot containing the vial; the long syringe needle; the lead shield with lead glass window surrounding the syringe; the tissue paper held so as to prevent splashes and drips; the double container for the volumetric flask; the large lead pot used for storage of the isotope before and after dispensing.

3.8.2 Contamination and the ingestion hazard

The spread of contamination is a most insidious hazard to experimental results. Inadvertent use of a contaminated flask or beaker during the transfer of a low activity sample can easily ruin an experiment which may have been going on for several months. If undetected it could lead to totally misleading conclusions and further waste of time or even the publication of erroneous information. The health hazard arising from contamination should not be minimized, but neither should it be exaggerated to the point where students are afraid to tackle tracer experiments where the dangers are minimal.

Contamination control boils down to good house-keeping and absolute cleanliness. Eating, drinking, smoking, application of cosmetics, are all banned in radiochemical laboratories. Laboratory surfaces should be hard and shiny, e.g. gloss painted walls and woodwork, melamine plastic bench surfaces, welded lino floors, stainless steel sinks, etc., all designed for easy washing and decontamination. Fume cupboards and air changes should be adequate for the level of work envisaged – further details are available in the publications cited in Appendix D.

The principle of containment is paramount. Experiments should as far as possible be conducted over large trays lined with absorbent paper. Within the tray, isotope containers should be inside more stable secondary containers, e.g. beakers. Transfer of liquids should be by syringe or bulb-operated pipettes; no mouth operations are permitted. Paper tissues should be widely used for wiping up droplets and for handling laboratory services such as taps, switches, monitors, etc., to prevent transfer of activity from one place to another. The tissues then become low level radioactive waste and should be immediately placed in a suitable container for disposal.

Laboratory coats should always be worn. Whether or not to wear gloves is often a 'grey area' for decision. They should certainly be worn during initial dispensing of an undiluted radiochemical. Thereafter for low or medium toxicity nuclides it becomes a matter for personal preference. Thin rubber or vinyl surgical gloves can certainly prevent contamination of the hands but they are uncomfortable to wear for extended periods, become extremely slippery if wet and often make simple operations like weighing more difficult. On the whole, for low level tracer work (<1 MBq, ~ 25 μCi) we prefer to work without gloves, but insist on frequent hand washing and a monitoring check before leaving the laboratory.

At the end of a particular experimental procedure, all apparatus which is known to be contaminated should be segregated and decontaminated at the earliest possible opportunity. Apparatus and areas which may, inadvertently, have become contaminated should be carefully monitored and decontaminated if necessary. Decontamination in general means giving the object in question a very thorough wash, using a good quality detergent and/or a suitable solvent. Various surfactant detergents are available commercially and these are excellent for decontaminating glassware and apparatus (see Appendix E). However, even using such powerful detergents repeated washing may be necessary, and if contamination persists then expert advice should be sought.

3.9 Storage, disposal and record keeping

The procedures to be adopted will usually be covered by local rules and regulations, but a few general points are worth noting. Good scientific practice requires that all the quantitative and descriptive details of an experiment should be recorded in a durable practical book. When using radioactive materials the account should also include details of the amounts and locations of the isotopes involved and their ultimate fate. It is then a fairly simple matter to prepare a 'balance sheet' of activity received and activity disposed of.

Obviously, unused or re-usable radioisotopes should be stored in clearly labelled containers in a secure place which is only accessible to authorized persons. Radioactive waste should not be stored any longer than is absolutely necessary and should be disposed of by the simplest legal route as soon as possible. The only exceptions to this would be storage of short lived nuclides until they have decayed to the point where they can be got rid of by normal refuse disposal procedures.

It is also necessary to keep records of radiation doses, contamination incidents, tests of monitoring instruments, medical reports, etc., but these will usually be dealt with by the appropriate safety authority of the institution. Full details of all these matters will be found in the publications in Appendix D.

4 Detection and measurement of radioactivity in biological materials

4.1 Introduction

Ionization is the property used to measure radioactivity. The four most important methods of detection involve (a) ionization produced in gases (ionization chambers and Geiger–Müller detectors), (b) ionization produced in solids (semiconductor detection), (c) scintillations produced in fluors (solid or liquid scintillation detectors) and (d) the blackening of photographic emulsions (autoradiography and particle track detectors). The first three methods require additional electronic equipment to amplify and analyse the tiny electrical pulses produced in the detectors. The amplified pulses are then recorded in one of two ways, either by integration or by individual pulse counting.

Integrating instruments include electrometers which measure the total amount of ionization produced, and ratemeters which give direct reading on a meter or pen recorder of an approximate mean count rate. The electronic circuitry is relatively simple and inexpensive. They are widely used in radiation protection for measuring dose rates or levels of contamination. Ratemeters are also used when an analogue display is required on, for example, a chart recorder from a detector which may be scanning a thin layer chromatogram or which may be following the uptake of an isotope by a specific organ in a living organism.

Individual pulse counting is always used where quantitative measurement of activity is required. The number of pulses recorded in unit time is directly related to the disintegration rate of the particular isotope via the counting efficiency, E (since E = observed net count rate ÷ disintegration rate) after suitable corrections have been applied for background radiation, etc. (see p. 77). Simple instruments usually incorporate a **scaler** which records the total number of pulses, a **timer** which records the counting time, and some facility for stopping the count either at a pre-set time or at a pre-set number of counts. More complicated instruments will incorporate circuits to analyse and sort pulses in terms of their size or energy (e.g. pulse height analysers) and may also incorporate extensive computing facilities as in the modern liquid scintillation counters which are essential for biological research using sulphur-35, carbon-14 or tritium.

Selection of a suitable means of detection for a particular experiment will depend upon the type (α, β or γ) and energy of the radiation, the amount of activity present, and the complexity of the sample (e.g. single or multiple labelling, live animal or purified metabolic product, etc.). The main attributes of the most important detectors are set out in the following sections.

4.2 Ionization chambers

Ionization chambers are made in all shapes and sizes from tiny pocket dosemeters to the large detectors used to monitor the environment

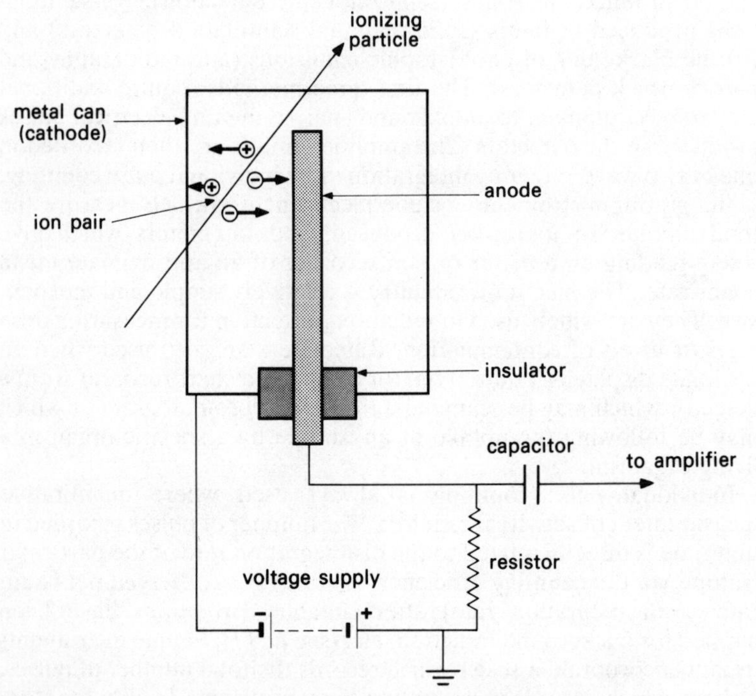

Figure 4.1 Cross-section through a simple air-filled ionization chamber. Similar devices are used extensively in radiation safety monitoring.

around nuclear reactors. In some cases they are designed so that a radioactive sample can be placed inside the sensitive volume to reduce the losses due to absorption of the radiations by intervening matter. An ion chamber consists of an anode and a cathode separated by a gas. Many gases have been used, depending upon the type of chamber. For health physics purposes air at atmospheric pressure is frequently satisfactory and in metabolic experiments air containing respired $^{14}CO_2$ can be used. The principle of operation is illustrated in Figure 4.1. As ionizing radiations pass through the sensitive volume ion pairs are created and are immediately attracted towards the charged electrodes. The dependence on the applied voltage is illustrated in Figure 4.2. If the applied potential is too low, i.e. less than

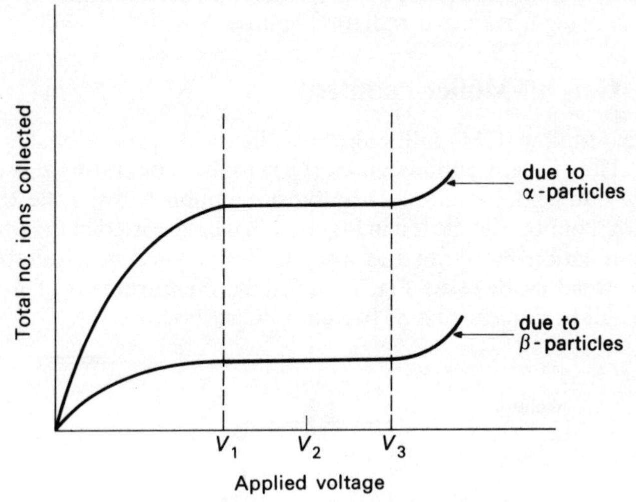

Figure 4.2 Characteristic curve for an ionization chamber showing the difference in response to α-particles and β-particles, which is due to the much greater amount of ionization produced by the heavier α-particles.

V_1, many of the ions will collide and recombine before reaching the electrodes. However, at voltages between V_1 and V_3 all the ions which are formed are collected and the chamber is normally operated at voltage V_2 so that voltage fluctuations do not affect its performance.

Above voltage V_3 the initially formed ions are accelerated by the applied potential to such an extent that on collision with gas molecules they cause further ionization resulting in the upturn shown. Here we enter the 'proportional' region and at higher voltages a region of 'limited proportionality' before finally entering the 'Geiger' region (see GM counters, next section). The proportional regions are of historical interest but are not used much for counting biological samples. There is a clear distinction between α-particles and β-particles due to their relative specific ionizations of ~4000 : 50 (p. 24) and thus ionization chambers may be used to differentiate between different types and energies of radiation (like scintillation detectors, where this topic is more fully discussed on p. 51). The magnitude of the capacitor and resistor determine the response time of the detector. If the response time is kept short (down to 1 μs), the ionization chamber can be used for individual particle counting. However, if the response time is long the device effectively measures the total ionization produced rather than individual particles, and in this mode is widely used in health physics to measure dose rates. Ionization chambers must be calibrated and re-calibrated from time to time using a standard radiation source.

4.3 Geiger–Müller counters

Geiger–Müller (GM) tubes also exist in a wide variety of shapes and sizes. They usually contain an inert gas (helium or argon) at pressures below atmospheric. One of the most common types is the thin end window counter illustrated in Figure 4.3, which is used in the probes of contamination monitors and also, when installed in a suitable lead shield (lead castle) (see Fig. 2.7), for the measurement of β-activity from solid samples such as barium [^{14}C]carbonate.

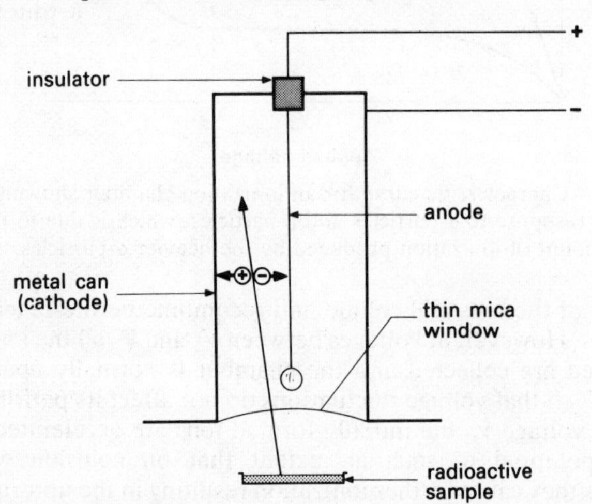

Figure 4.3 Cross-section through a typical thin end window Geiger–Müller tube of the type which is in wide use for monitoring contamination and for measuring weak β-particles.

Another fairly common type is the windowless flow counter used to scan thin layer and paper chromatograms. As there is no window the sensitive volume is usually partly enclosed, except for a narrow slit which is placed very close to the chromatogram and through which the β-particles pass. The correct gas mixture is maintained by allowing it to flow continuously into the counter and out through the slit. The width of the slit determines the sensitivity and resolution of the counter, the two parameters being mutually exclusive. GM counters operate in a higher voltage range than ionization chambers. Thus, when a particle enters the sensitive volume and forms ion pairs, these are accelerated to such an extent that they acquire enough energy to cause further ionization as they collide with other gas molecules. The secondary ions so formed also accelerate and cause further ionization, the net result being an avalanche of ions producing a large pulse from the counter. This process, known as gas amplification, can occur over a voltage range which gives the GM counter a characteristic curve as shown in Figure 4.4. As in the case of the ionization chamber the operating voltage, V_2, is selected in the middle of the plateau. There is an important difference between the two types of counter; gas amplification in the GM counter means that the final pulse from the counter is independent of the initial number

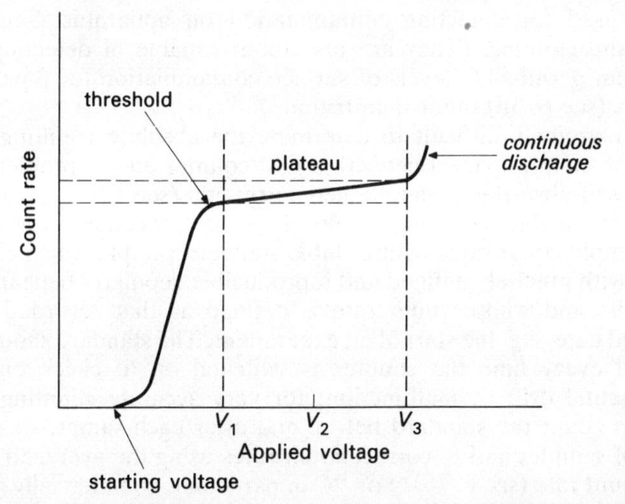

Figure 4.4 Response curve for a Geiger–Müller counter. Counters are characterized by the threshold voltage, length and slope of the plateau, dead-time and temperature range for operation. If the counter is operated at voltage V_2, fluctuations due to mains voltage variation will have only minor effects on the observed counting rate.

of primary ions and thus, unlike the ionization chamber, the GM counter is unable to differentiate between different types and energies of radiation. However, pulses from a GM counter are so large (gas amplification factors of up to 10^{10} are possible) that the subsequent electronic amplification can be very simple and cheap.

Two other problems with GM counters arise from the large numbers of positive ions which are formed. The positive ions are slow moving relative to electrons and may take as long as 200 μs to clear away. During this time the counter is 'dead', i.e. unable to record particles entering the sensitive volume. Thus, it becomes necessary to correct for lost counts (see p. 77) and GM counters are unsuitable for samples with high count rates ($>100\ s^{-1}$) since the correction becomes statistically unacceptable. Eventually the positive ions strike the cathode and at this point may cause release of further electrons which could re-trigger the whole avalanche process. This undesirable effect is prevented by incorporating a quenching agent (usually a trace of bromine vapour) into the counter gas filling which assists in energy dissipation without causing further ionization.

Thin end window GM counters are capable of detecting carbon-14 β-particles with an efficiency of 1–5%. They are incapable of detecting tritium at all. However, despite the various disadvantages mentioned above they are extremely robust, reliable and portable and are widely used for detecting contamination on apparatus, benches, hands and clothing. (They are just about capable of detecting the maximum permissible levels of surface contamination for β-particle emitters (see p. 36) other than tritium.)

It is extremely difficult to determine the absolute counting efficiency of samples placed beneath a GM counter due to problems of sample self-absorption, sample geometry, etc. (see Ch. 5). For precise, reproducible, counting results the usual procedure is to compare sample count rates with a stable standard sample which can be placed with precisely defined and reproducible geometry beneath the GM tube and whose count rate is defined as that recorded at a specified date, e.g. the start of an experiment. The standard should be counted every time the counter is switched on to check on any instrumental drift or malfunction; for very accurate counting it is usual to count the standard before and after each sample or small batch of samples and to correct count rates using the averaged standard count rate (see p. 78). For ^{14}C in particular, commercially available discs of ^{14}C-labelled poly(methyl methacrylate) make very reliable reference standards.

4.4 Semiconductor detectors

The most recent developments in radiation detection involve the use of semiconductor materials such as silicon and germanium. These detectors may be crudely described as solid state versions of gas ionization chambers with the considerable virtues of much greater stopping power for the radiation and very much higher resolution of energy peaks. They are widely used in nuclear research and activation analysis (see p. 107) but as yet their prophesied impact in the biological sciences and tracer work has not materialized so that they are not considered further here.

4.5 Scintillation detectors

These detectors may be subdivided into solid and liquid scintillation counters. The former are used to measure γ-radiation and the latter have been specially developed for the measurement of weak or very weak β-particle radiation. The fundamental physical processes are somewhat different in the two types of counter but the general features have much in common. Thus, passage of radiation through matter results in deposition of energy by ionization and excitation (see p. 22). Excitation results in light emission and if the radiation is passing through a substance which fluoresces and is also transparent to the light emitted, some of the energy used for ionization can be converted by energy transfer to give further excitation and increased light emission. Such substances, which emit tiny flashes of light when bombarded with particles, are termed **scintillators** or fluors. Becquerel first observed such behaviour in 1899 and Rutherford and Geiger later devised a method for counting individual α-particles by visual observation of flashes of light on a zinc sulphide screen. In modern instruments the eye is replaced by one or two **photomultiplier tubes** which are capable of detecting extremely small light flashes with resolving times (dead times) so small that counting losses are insignificant for all normal assay samples. A schematic diagram of a scintillation detector is given in Figure 4.5.

As radiation passes through the scintillator it generates light photons. A proportion of photons strike the photocathode causing the emission of a proportional number of electrons which are attracted to the positively charged first dynode. Each electron striking the first dynode liberates several secondary electrons which are then accelerated by a higher positive potential to the second dynode where further electron multiplication occurs, and so on down eleven or thirteen dynode stages of ever-increasing potential. A single electron from the photocathode can result in a pulse of 10^6–10^7 electrons

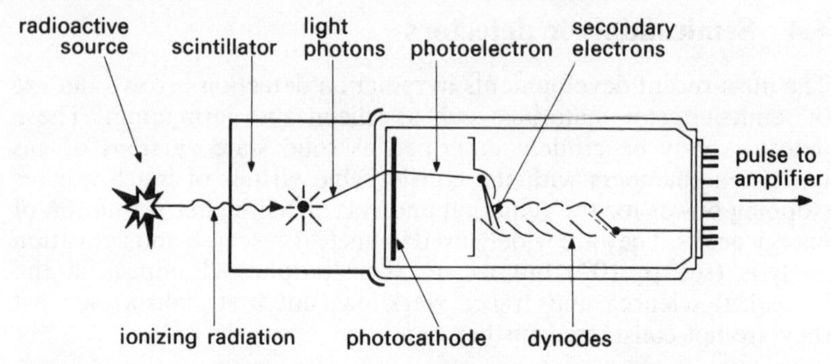

Figure 4.5 A scintillation detector incorporating a photomultiplier tube with linear focused dynodes for a fast response time. When light photons strike the photocathode, electrons are released and accelerated by a potential difference toward the first dynode. Each electron striking the first dynode causes the release of two or more electrons which are then accelerated to the second dynode – and so on down the multiplier chain.

at the output from the electron multiplier, which can then be further amplified and used to operate counting equipment. A disadvantage of electron multiplication is that stray electrons within the photomultiplier tube can give rise to spurious pulses unrelated to the original radioactivity so that the random background count rate tends to be high. This problem can be partly alleviated by cooling the photomultiplier, but the best solution is to have two photomultiplier tubes viewing the scintillator with their outputs fed into a **coincidence circuit** where random electron pulses from either tube are rejected and only those pulses resulting from an event seen by both tubes simultaneously is passed on to the counting circuits (see p. 56) In this way low stable background count rates are readily obtained.

4.5.1 Integral or differential counting modes

An important feature of scintillation detectors is that, like ionization chambers, they are energy sensitive; that is, the size of the pulse generated is directly proportional to the energy of the radiation deposited in the scintillator. Thus it is possible to use electronic circuits to 'sort out' or discriminate against pulses representing radiations of different energy. A simple counter is represented in Figure 4.6. Such a counter is used for **integral counting**, where the **discriminator** is set to reject all low level (mainly background) pulses below a certain value but all other pulses from the detector are counted regardless of their size. The arrangement is useful for measuring the count rate of a sample but a great deal of potentially useful information is disregarded.

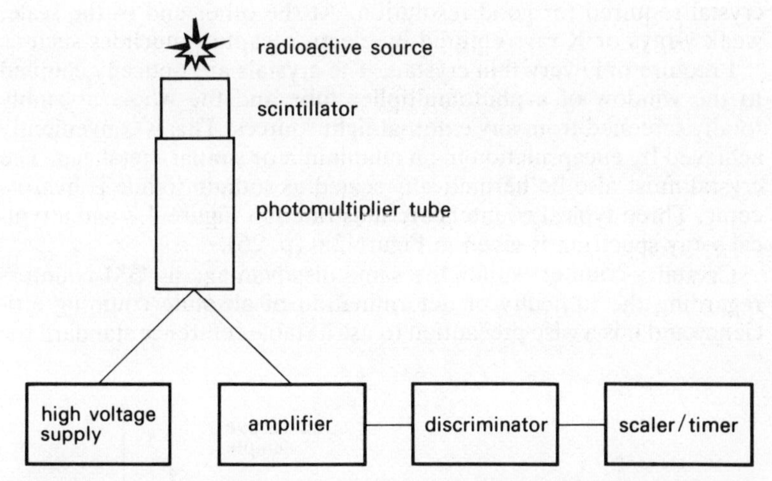

Figure 4.6 Block diagram of a simple scintillation counter. Although shown as separate units, the electronic circuits are frequently assembled in one cabinet.

If the discriminator in Figure 4.6 is replaced by a **pulse height analyser**, the scintillation counter can be used as a spectrometer to determine the energies of different radiations. The simplest pulse height analyser consists of two discriminators, one which rejects all pulses below a specified level and one which rejects pulses above a specified level. Only those pulses falling in the **window** between the two levels are counted. The system is referred to as single channel pulse height analysis and counting in this mode between specified limits is known as **differential counting**.

The energy spectrum of a particular radiation may be determined in two ways. Either the spectrum is scanned by gradually moving a very narrow window of a single channel analyser over the required energy range or by using a multi-channel pulse height analyser where a large number of very narrow channels (typically 1024, 2048 or 4096) accumulate pulses continuously over the entire energy range.

4.5.2 Solid crystal scintillation counters

There are several types of organic and inorganic crystals which behave as scintillators. The most commonly used material is sodium iodide which contains a small amount of thallium as an activator. This material can be produced in large, transparent crystals and because of its high density it has a high stopping power and is ideal for counting γ-rays. Generally, the higher the energy of the γ-rays the larger the crystal required for good resolution. At the other end of the scale, weak γ-rays or X-rays emitted by electron capture nuclides such as ^{125}I require only very thin crystals. The crystals are optically coupled to the window of a photomultiplier tube and the whole assembly totally screened from any external light sources. This is conveniently achieved by encapsulation in an aluminium or similar metal can. The crystal must also be hermetically sealed as sodium iodide is hygroscopic. Three typical counters are illustrated in Figure 4.7 and a typical γ-ray spectrum is given in Figure 2.6 (p. 26).

Crystal γ-counters suffer the same disadvantage as GM counters regarding the difficulty of determination of absolute counting efficiency and it is a wise precaution to use a stable reference standard for

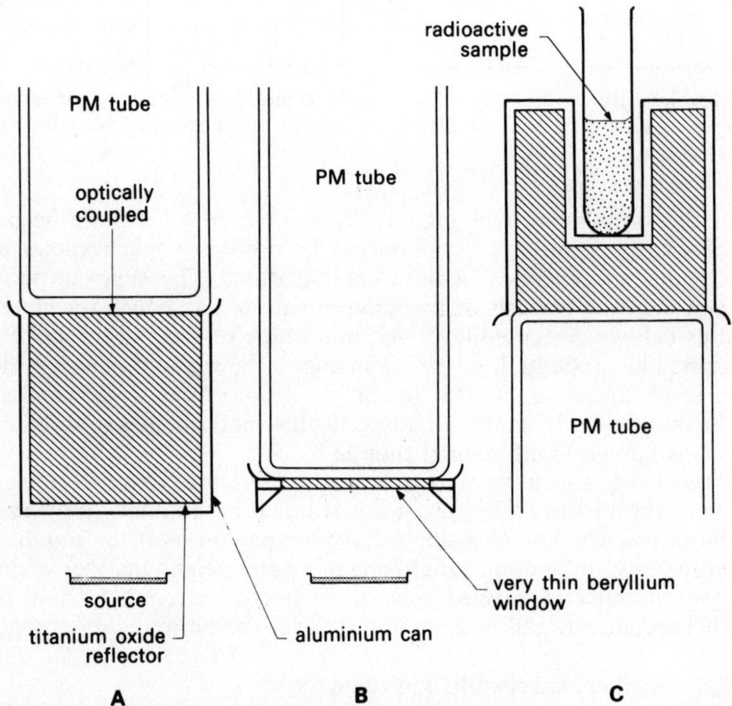

Figure 4.7 Examples of scintillation detectors incorporating thallium-activated sodium iodide crystals. Type A consists of a large cylindrical crystal (typically 75 mm × 75 mm) of the type used for γ-ray spectroscopy. Type B has a very thin window and is suitable for measuring very weak γ-rays or X-rays such as those emitted by iodine-125. Type C is the popular 'well-type' for routine activity measurements of γ-emitters of all types.

Table 4.1 Count rates obtained from identical one millilitre samples of an aqueous solution of iodine-125 using standard 20 ml counting vials in a sodium iodide 'well-type' crystal viewed by two photomultiplier tubes.

Vial type	*Added potassium iodide carrier (% w/v)*	*Observed count rate (min^{-1})*
plastic (manufacturer no. 1)	–	69 483
plastic (manufacturer no. 2)	–	69 232
plastic (manufacturer no. 2)	10	59 509
plastic (manufacturer no. 2)	20	52 267
glass (manufacturer no. 1)	–	48 443
glass (manufacturer no. 2)	–	53 438
2 ml mini vial insert inside standard plastic vial	–	71 792
2 ml mini vial insert inside standard glass vial	–	51 814

checking instrument performance from time to time. Modern counters, however, are very stable and where large numbers of closely similar samples are being counted, e.g. in radioimmunoassay, it is reasonable to assume a constant (although unknown) counting efficiency during the course of the experiment.

For emitters of very weak γ-rays and X-rays it is most unwise to assume constant counting efficiency if the sample container or sample composition is variable. The results in Table 4.1 clearly illustrate this point. As there is no satisfactory method of correction for such variable behaviour an alternative counting method such as liquid scintillation counting should be considered.

4.5.3 Liquid scintillation counters

The development of the modern liquid scintillation spectrometer has probably been as important to medical and biological researchers as the development of infrared or nuclear magnetic resonance spectrometers was to chemists. There is fierce competition between the different manufacturers and this has resulted in the development of extremely efficient, adaptable and reliable instruments which are capable of dealing with large numbers of samples and of carrying out all the necessary computation, manipulation and tabulation of data. The literature on the subject is vast (many excellent books and monographs are listed in Appendix D) and folk-lore abounds. In this section we attempt to summarize the most important aspects of the technique.

(a) The sample In liquid scintillation counting the radioactive sample is dissolved or suspended in a liquid scintillant which in some cases can be a quite complex mixture. The total volume of scintillant plus sample can vary from around 2–18 cm^3 and is usually contained in a glass or plastic vial. Sample preparation is dealt with in the next section but samples usually contain three components in addition to the radioactive substance. The **primary solvent** has to be an aromatic solvent most usually toluene or a mixture of xylenes. The **primary solute** is also aromatic in character but usually contains several rings which may include heterocyclic structures. The π-electron systems of these compounds are essential for the scintillation process. Many compounds have been described, but the only two in wide use are PPO (**I**) and butyl-PBD (**II**). The latter compound gives solutions which are more stable to light and more resistant to quenching (see p. 55) but more susceptible to decomposition in alkaline conditions.

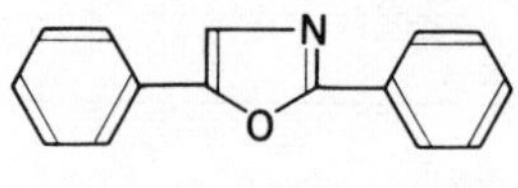

PPO
(2,5-diphenyloxazole)

I

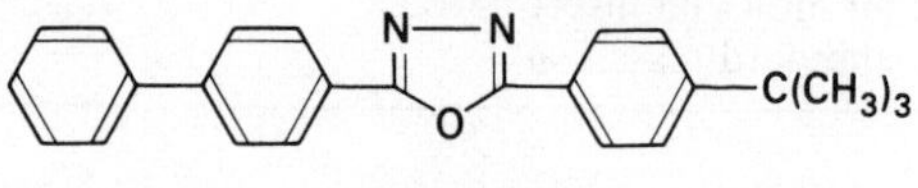

Butyl-PBD
(2-(4′-*t*-butylphenyl)-5-(4″-biphenylyl)-1,3,4-oxadiazole)

II

Most of the scintillant recipes recorded in the literature include secondary solutes such as POPOP, dimethyl-POPOP or bis-MSB. They were originally required as 'wavelength shifters' to match the wavelength of the light output to the older type of photomultiplier tubes. Modern tubes have such a wide spectral response, however, that use of secondary solutes is no longer necessary and can even lead to reduced counting efficiencies – they should be omitted, if only on grounds of economy.

An 'ideal' sample would consist of a radioactive aromatic hydrocarbon (^{3}H- or ^{14}C-labelled toluenes are frequently used as reference materials) blended with a liquid scintillant composed of primary solvent containing 0.3–1% of primary solute. β-particles pass through the solution causing ionization and excitation of the solvent molecules. A small fraction of the deposited energy is transferred and excites the solute molecules which emit light photons as they return to their ground states. The photons are then observed by photomultiplier tubes as described in Section 4.5.7 (p. 49). The enormous advantage that this counting method has over those described previously is that all the disintegrations occur within the sensitive volume and problems due to sample self-absorption, window thickness, backscatter, etc., do not arise. The 'ideal' samples would give counting efficiencies of around 95% for carbon-14 and 60% for tritium. The lower value for tritium is due to the extremely low energies of the β-particles, many of which cannot produce a sufficient number of photons to activate the photocathodes of the photomultipliers.

Ideal samples rarely occur in practice. Many radiochemicals and most radioactive biological samples are insoluble in toluene or xylene and hence it is necessary to devise means of blending or suspending the sample with the scintillant. Often a **secondary solvent** or a **solubilizer** is necessary and further details are given in the next section. However, an unfortunate consequence is that almost any substance which is added to the 'ideal' scintillant will reduce the scintillation efficiency (i.e. the number of photons emitted per particle) by competing with the primary solute for energy transfer. Such behaviour is referred to as **impurity quenching** and can vary from very mild effects with substances such as methanol or ethanol to extremely severe effects with substances such as chloroform and carbon tetrachloride which can reduce counting efficiencies to zero in some instances. Coloured materials also cause problems due to **colour quenching** where the emitted light is absorbed within the sample and is thus prevented from reaching the photomultipliers. A further complication which sometimes arises is the production of spurious light pulses from **fluorescence** or from **chemiluminescence** within the sample.

Variable counting efficiency is the biggest drawback in liquid scintillation counting but fortunately several reliable methods are available to correct for these variations (see p. 57).

(b) The instrument A modern spectrometer will incorporate most of the features illustrated in Figure 4.8, although different models will emphasize different features and may be more or less complicated. Most models incorporate automatic sample changers which hold around 100–500 counting samples in either serpentine belts or individual trays. Two photomultiplier tubes are quite standard and are operated through a coincidence circuit to reduce the background count (see p. 50). Coincident pulses from the two tubes are added together in the pulse summation circuit in order to provide larger pulses which are independent of the location of the original scintillation event within the sample. This in turn leads to better resolution of overlapping β-spectra in double labelled samples (see p. 110). The amplifiers can have either a linear or logarithmic response. Each has its own particular virtues but discussion is outside the scope of this book. Linear amplifiers require attenuators while logarithmic amplifiers do not since they cover the entire pulse amplitude range. Most instruments incorporate two or three separate channels of pulse height analysis for sample counting plus two further channels for use

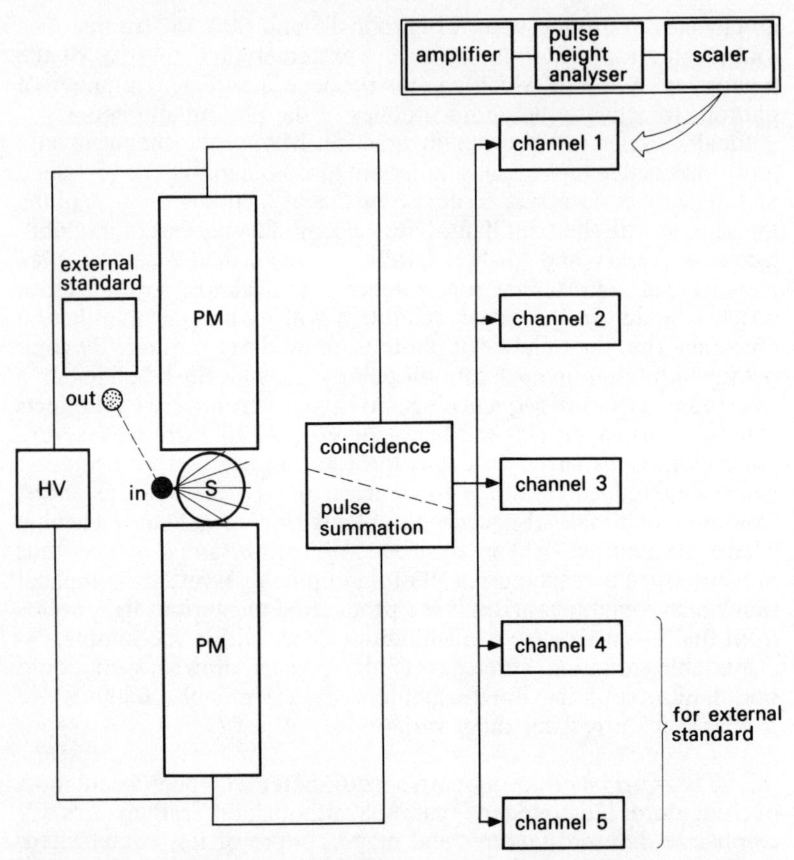

Figure 4.8 Block diagram of a modern liquid scintillation counter. Events in the scintillator S which are seen simultaneously by both photomultiplier tubes give rise to pulses which are passed by the coincidence circuit and added together in pulse summation. Random, non-coincident pulses are rejected. Summed pulses are then passed simultaneously to up to five separate channels identical to that shown for channel 1. The pulse height analysers can be set to sort pulses of different energies for quench monitoring (see p. 57) or double isotope counting (see p. 110).

with the external standard (see p. 60). Channels may either be selected manually or in many cases are on plug-in modules pre-set by the manufacturer for counting a particular nuclide. The most recently developed instruments have multi-channel analysers scanning the entire relevant energy spectrum and the selection and adjustment of all parameters is controlled automatically by microprocessors. The output from each channel is fed either to scalers or to memory core where the total counts are recorded. All the accumulated data can then be fed to computational devices of various complexity for final data reduction.

(c) Quench correction Since almost anything added to a liquid scintillant, including the radioactive sample itself, can cause a reduction in counting efficiency, it is necessary to have some mcans of monitoring the variations which occur. The three most important methods are described below. Other methods are discussed in the texts listed in Chapter 7 or in manufacturers' literature.

(i) *Internal standardization* In this method a sample is counted and then 'spiked' with a known amount of the same isotope and re-counted. The increase in the count rate is a measure of the counting efficiency, E:

$$E = \frac{\text{net (sample + standard) count rate} - \text{net sample count rate}}{\text{disintegration rate of standard}}$$

The substance used as an internal standard should have a high specific activity so that only a very small amount need be added, thus not materially altering the counting sample, and should be chemically stable, involatile and non-quenching. For ^{14}C and ^{3}H the labelled hexadecanes are ideal, for ^{35}S dioctyl sulphide can be used and for ^{125}I iodobenzene is suitable.

The method is of fundamental importance since it is the means by which the two other methods to be described are calibrated. Its advantages include simplicity coupled with accuracy and it is suitable for heavily quenched samples. The disadvantages are that it is tedious, every sample must be opened, spiked and re-counted (even more time-consuming with double-labelled samples) and, once spiked, a sample cannot be recovered or re-checked at a later date.

(ii) *Sample channels ratio* The importance of the continuous energy spectra of β-particle emitters was mentioned in Chapter 2 (p. 25). In a liquid scintillation counter, β-emitters give rise to closely related pulse height spectra which can be observed if the relevant energy range is scanned with a very narrow single channel or a multi-channel analyser. If an unquenched sample is scanned then quenched to give a lower efficiency and re-scanned, a definite shift of the entire spectrum towards lower energies is observed, as illustrated in Figure 4.9. If two channels of pulse height analysis are set as shown in Figure 4.9, then any shift in the spectrum (i.e. any change in E) can be detected by observing the ratio of the counts in the two channels. The positioning and width of the 'monitoring channel' (channel 2) in Figure 4.9

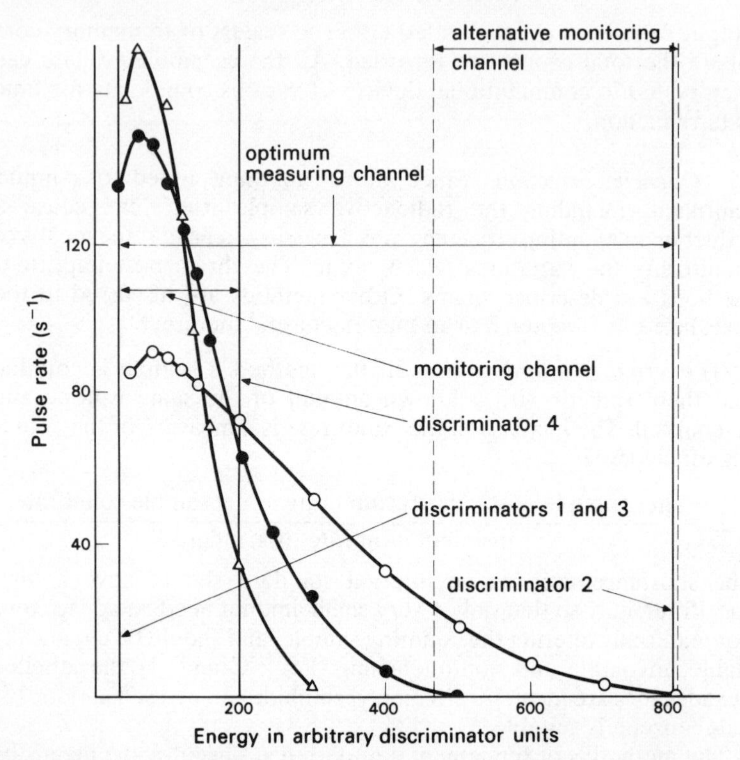

Figure 4.9 Pulse height spectra for ^{14}C (linear amplification) showing the shift to lower energies caused by impurity (acetone) quenching. Each point was obtained using a pulse height analyser with a narrow window of 50 discriminator units. ○, Unquenched; ●, 0.5% acetone; △, 2% acetone.

depends upon circumstances and particularly upon the range of quenching which is expected. In some circumstances, for low amounts of quenching, it may be positioned at the top end of the energy range as shown by the broken lines. However, for general purpose counting it is best placed as shown with upper discriminator 4 set at about one-third of the maximum for unquenched samples. The method requires calibration using internal standards. A series of vials is prepared, each vial containing exactly the same amount of radioactive standard and the same amount of scintillant but differing amounts of quench agent. The samples are then counted to a statistically satisfactory level (see p. 80) and a calibration curve prepared by plotting sample channels ratio,

$$R = \frac{\text{net counts in channel 2}}{\text{net counts in channel 1}}$$

against efficiency,

$$E = \frac{\text{net count rate in channel 1}}{\text{disintegration rate of standard}}$$

as shown in Figure 4.10.

Unknown samples can then be counted, their channels ratios computed and counting efficiencies determined by reference to the graph. The whole procedure can easily be programmed into a computer and the instrument will then print out the number of disintegrations per minute directly.

The advantages of this method are that only one count is required, accuracy is good, the sample is not contaminated (so that it may be re-counted or recovered) and the whole procedure can be programmed on a computer. The disadvantages are that it is not suitable for low activity samples (because of problems with background and counting statistics) and unless there is a very low level of quenching a separate curve is required for colour-quenched samples.

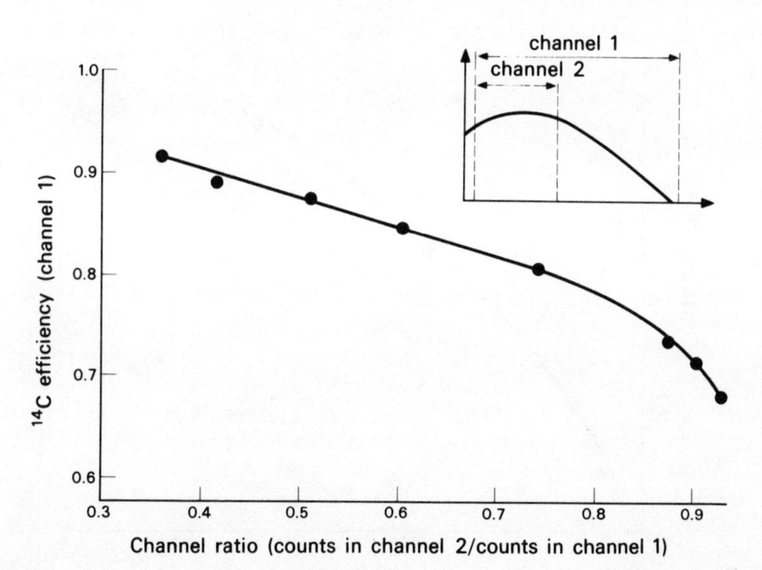

Figure 4.10 A quench correction curve prepared using a set of quenched standards and the channels ratio method.

(iii) *External standard channels ratio* Most modern counters incorporate an external standard which consists of a small pellet containing a γ-emitting nuclide such as ^{226}Ra, ^{137}Cs or ^{133}Ba and which is stationed in a shielded position remote from the counting chamber. On command, the standard can be transported mechanically or pneumatically to a precise location in the counting chamber where the γ-rays will irradiate the counting vial and its contents. A rather diffuse γ-ray spectrum can be observed but in particular the Compton edge (see Ch. 2, p. 26) can be seen to shift to lower energies with increasing amounts of quenching. If two channels are set to encompass that part of the energy spectrum and to divide it in much the same way as for the sample spectrum in the previous method, then an external standard channels ratio can be obtained and used as a measure of the extent of quenching. As before, a set of quenched standards is required in order to set up a calibration curve (Fig. 4.11). Thereafter the computation of results is much the same as for method (ii).

The method is slightly less reproducible than the sample channels ratio and counting times are slightly longer since the external standard requires a separate, although relatively short (~1 min), counting period. Its main advantage is that it can be used for low activity

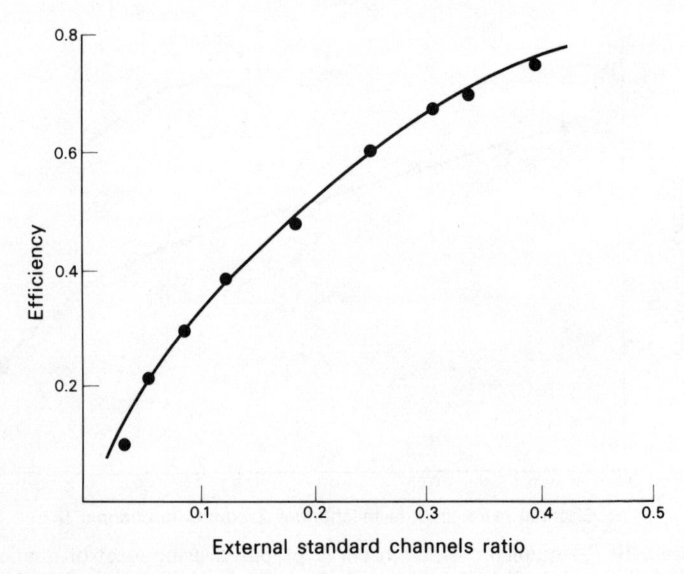

Figure 4.11 A typical quench correction curve using the external standard channels ratio method.

Table 4.2 Cerenkov counting efficiencies for various nuclides.

Isotope	E_{max} *(MeV)*	*Approximate counting efficiency (%)*
^{36}Cl	0.71	5
^{40}K	1.32	34
^{24}Na	1.39	40
^{89}Sr	1.46	41
^{32}P	1.71	50

samples with count rates near to background – i.e. the kind of sample which often is the end product of a biological experiment.

(d) Cerenkov counting For β-particles whose maximum energy exceeds about 0.7 MeV there is no need to bother with scintillant mixtures at all. **Cerenkov radiation** is light which is emitted when a charged particle passes through a transparent medium with a speed greater than the speed of light in that medium. (A simple analogy is the 'sonic boom' produced by an aircraft when its velocity exceeds the speed of sound in its surrounding air.) Thus, when β-particles of sufficient energy pass through water, photons of light are generated and can be detected by the photomultipliers of a liquid scintillation counter. The photon yield is similar to that produced by tritium in a liquid scintillation solution, so that with the counter settings as for tritium colourless aqueous solutions may be counted directly. A few of the nuclides which may be counted in this way are listed in Table 4.2. Samples are not quenched by colourless additives but colour quenching can occur and can be corrected for by a samples channel ratio method. This method is very simple, requires no expensive scintillant and is well worth considering for the more energetic nuclides. It is also useful in double isotope counting (see p. 114).

4.6 Sample preparation in liquid scintillation counting

4.6.1 General principles

Most measurements of radioactivity in biological samples are made using liquid scintillation counting. For this to be successful it is very important that considerable attention is given to sample preparation because photon emissions from a counting vial are critically dependent upon the molecular environment within the vial and no two samples are ever completely alike. Therefore, in order to obtain an acceptable degree of accuracy and reproducibility, it is necessary to ensure that optimum conditions are established and maintained for each type of sample.

Maximum counting efficiency is attained when the radioactive sample is uniformly dispersed throughout the scintillation medium (either as a true solution or as a dispersion of particles or droplets (whose diameter is small with respect to the path length of the α- or β-particle)) and under these conditions the ideal 4π counting geometry is approached. However, very many biologically derived samples are hydrophobic in nature or are solids and do not readily dissolve in the solvents such as toluene usually employed for scintillant solutions. Thus an important part of sample preparation concerns the method used to render the radioactive material soluble or miscible with the scintillant in such a way that, where possible, a homogeneous preparation is produced and that reasonably large amounts of sample are incorporated into the counting mixture with the minimum amount of quenching.

There is an extensive literature on sample preparation, and for the most part well tried methods exist for almost every type of sample. In the account that follows we have confined our attention to those methods most commonly employed in our laboratories, but the reader should be aware that other methods described elsewhere might be equally acceptable. The most important requirement is that when a new sample preparation method is used it should be completely standardized for sample–scintillant mix and counter using the procedures described in Section 4.5.3.

4.6.2 Non-aqueous samples

These generally present no problem. Such samples (e.g. lipid material) are soluble in non-polar solvents and can be incorporated directly into the scintillant mixture. The only limitations concern the miscibility of the sample solution with the scintillant mixture and also the degree of quenching which either sample or sample solvent produces.

4.6.3 Aqueous samples

Many biological samples exist as aqueous solutions (e.g. nucleic acids, amino acids, sugars, etc., very often dissolved in buffers or salt solutions). Since the solubility of these solutions is very low in solvents such as toluene, the scintillant mixtures have to be modified in order to allow the aqueous sample to be incorporated. Two methods are commonly employed.

(a) Mixtures producing homogeneous solutions These involve the dilution of the scintillant by a more polar solvent such as ethanol in order to permit the addition of limited amounts of water. For sample volumes of up to 0.2 cm^3 a simple mixture of toluene (seven parts) and ethanol (three parts) together with either butyl-PBD (0.8% for ^3H or 0.5% for ^{14}C) or PPO (0.3–0.5%) is satisfactory for many aqueous solutions, although not adequate for those containing a high proportion of salts. For larger volumes of aqueous solutions a mixture such as Bray's solution can be used. This consists of

naphthalene	60 g
PPO	4 g
POPOP	0.2 g (can be omitted)
methanol	100 cm^3
ethylene glycol	20 cm^3
1,4-dioxan	to make 1 dm^3

Bray's solution can hold up to 30% of water and gives reasonable counting efficiencies; however, its use has been largely superseded by the easy availability of commercial mixtures designed to allow the incorporation of relatively large volumes of aqueous solutions for counting (see next section) without the associated disadvantages such as the peroxidation of dioxan that are inherent in using Bray's solution.

(b) Mixtures producing colloidal solutions Colloidal or emulsion counting systems based upon mixtures containing surfactant agents such as Triton X-100 possess high counting efficiencies and a high capacity for dissolving aqueous samples, including those containing large amounts of ionic substances. Their disadvantage is that they are thermodynamically unstable and tend to undergo abrupt changes in appearance and quality with only minor alterations in composition or temperature. Most emulsion counting systems show four regions of heterogeneity when increasing amounts of water are added, from a region of slight opalescence (in some cases this may not be apparent) to clear, to two separate phases and then to a gel. High efficiencies of counting are given both in the clear phase and in the gel phase, but it is important to establish the stability (both short and long term) for any new combination of surfactant–scintillant–sample mixture that is tried. Table 4.3 shows the recommended composition of Triton X-100 : toluene mixtures for optimal counting of different aqueous samples of biological interest.

As well as Triton X-100-based mixtures a number of commercial products are available. The characteristics and capabilities of these are well described in the relevant manufacturers' literature and the

Table 4.3 Recommended compositions of Triton X-100:toluene mixtures for optimal counting of different aqueous samples of biological interest.

Sample	*Scintillant composition, Triton X-100:toluene (v/v)*	*PPO*† *(dm^3)*	*Counting mixture*		*Merit value,*‡ *MIV*
			Scintillant (cm^3)	*Sample (cm^3)*	
water	1 : 1	8	6	4	1231
8 M urea	1 : 1	8	6	4	1142
5% sucrose	2 : 3	6	5	5	989
2 M NaCl	7 : 3	8	7	3	989
ammonium formate (0.03 M)	2 : 3	5	5	5	778
ammonium formate (1.0 M)	3 : 4	8	7	3	736
TCA (5%)	13 : 7	10	8.5	1.5	662
PCA (5%)	3 : 1	3	6	4	1148
formic acid (0.1 N)	6 : 11	10	8.5	1.5	706
HCl (1.0 N)	2 : 5	8	7	3	1030
HCl (3.0 N)	5 : 11	5	8	2	748
Fischer's medium (+20% horse serum)	7 : 9	6	8	2	456
tryptone:yeast glucose (TYG)	1 : 1	10	8	2	536
nutrient broth	1 : 1	4.5	6	4	448
Eagle's MEM	1 : 1	3.3	8	2	313
cow's milk	3 : 5	3.3	8	2	664
human urine	1 : 1	4.5	6	4	963
human plasma	2 : 7	6	9	1	388

†The concentrations of PPO are not optimized for maximum efficiency in all cases.

‡The merit value (MIV) is standardized for the instrument efficiency and is equivalent to

$$\frac{\text{\% by volume of aqueous solution in sample} \times \text{\% counting efficiency} \times 100}{\text{\% counting efficiency of instrument reference standard}}$$

Data taken from a lecture by B. W. Fox presented at a Workshop in Liquid Scintillation Counting, Queen Elizabeth College, London, 1979.

reader is referred to the list of addresses in Appendix E should additional information be required.

When heterogeneous mixtures are counted care must be taken to apply the appropriate quench correction. If internal standardization is used, it is essential that the standard should be dissolved in the same phase as the original activity. This also applies to the preparation of standards for channels ratio quench correction curves. For most samples, sample channels ratio methods are best and give reasonably reproducible data. External standard ratio quench correction curves are usually unsatisfactory. However, if it is desired to use external standard methods, then the external standard ratio must first be matched with the sample channels ratio. In fact, if the sample channels ratio is plotted on a graph against the external standard channels ratio, all homogeneous counting samples should fall on one line. Marked deviation from this line is a good indication of heterogeneity.

4.6.4 Samples requiring the use of a tissue solubilizer

Very often it is necessary to measure the amount of radioactivity in plant or animal tissue. For this to be possible the sample must first be dissolved in a suitable solvent before it can be incorporated into a homogeneous counting solution.

Tissue samples can be solubilized by treatment with solutions of NaOH or KOH, but these reagents induce a high level of chemiluminescence in the scintillant mixture and also, by their very nature, limit the total amount of sample that can be incorporated into the mixture for counting.

More commonly used now are solubilizers containing quaternary ammonium bases which have very strong solubilizing powers but induce the minimum of chemiluminescence. One base frequently used is hyamine hydroxide, (*p*-(di-isobutylcresoxyethoxyethyl)dimethylbenzylammonium hydroxide), but this is gradually being superseded by other compounds with more suitable properties most of which are available commercially under various proprietary names (Table 4.4).

Table 4.4 A selection of commercially available tissue solubilizers.

Digestin	E. Merck, Darmstadt
Eastman Tissue Solubilizer	Eastman Organic Chemicals
NCS	Amersham-Searle Corporation
Protosol	New England Nuclear (NEN Chemicals GmbH)
Soluene 100/350	Packard Instrument Company
Tissue Solubilizer TS1/TS2	Koch-Light Laboratories Ltd

The quaternary ammonium base solubilizers are able to dissolve a wide range of biologically derived materials, including blood, protein, nucleic acids and various coarsely ground tissues, but in general they are unable completely to dissolve tissues containing bone, cartilage and collagen. Dissolution of material normally requires frequent agitation or swirling of the sample and is accelerated by heating up to 50°C. Temperatures above 50°C should be avoided since this can cause darkening of the solution and hence increase quenching. Where highly coloured solutions do become a problem it is possible to decolourize them by bleaching with hydrogen peroxide or benzoyl peroxide. However, these reagents may increase the amount of chemiluminescence within the sample and more importantly can cause loss of radioactivity through the loss of gases during the oxidation process. It is therefore better to avoid their use where possible and, if necessary, to employ alternative means of preparation for counting. One further point to note concerning the use of tissue solubilizers is the effect that these highly alkaline reagents have upon the stability of the scintillants. Tissue solubilizers should not be used in conjunction with butyl-PBD since this is unstable in alkaline solution; however, PPO is less affected and can therefore be used.

4.6.5 Samples requiring complete oxidation

Occasionally dissolution of a sample may produce a highly coloured solution which causes severe quenching or the radioactive content of the sample may be so low that too large a mass is required for adequate amounts of radioactivity to be measured. In these cases direct solubilization techniques are unsuitable and an alternative method is necessary. One such alternative is the direct oxidation of the sample followed by trapping and counting of the oxidation products. A very simple and convenient method of sample oxidation is Schöniger oxygen flask combustion. In this method the specimen (5–500 mg) is wrapped in a cellulose (filter paper) bag and burnt in an atmosphere of oxygen. Ignition is usually achieved by means of an electric current or spark and following complete combustion the gaseous products are trapped and subsequently counted. Commercially available 'sample oxidizers' using this method are common and have the advantage that the process is completely automated. In addition they are usually designed for the combustion of samples containing ^{3}H- and ^{14}C- and allow for the combustion products ($^{3}H_2O$ and $^{14}CO_2$) to be trapped and measured separately, which obviates the need for dual labelled counting (see p. 110).

4.6.6 Additional methods commonly employed for biological samples

(a) Counting of material separated on TLC plates In the absence of a special counter designed to scan TLC plates (see p. 116 radio-

active material must be removed from the plates and incorporated into the scintillation mixture in some way. The simplest and most direct way is to scrape the solid material from the plate and incorporate the solid particles into the scintillation mixture directly. Provided that equivalent amounts of TLC plate material are added to each scintillation vial (for example 1 cm^2 area from a 200 μm thick plate) then the relative amounts of radioactivity in each sample can be determined. However, it is not easy to determine absolute activities in this way because of the problems of self-absorption by the solid particles and because of the difficulty of obtaining uniform suspensions within the vials. The latter problem can to some extent be overcome by the use of hydrofluoric acid to dissolve the surface of the silica gel. A TLC spot is scraped off and transferred to a counting vial, ethanol (four drops) is added to soften the gel, then hydrofluoric acid (0.4 cm^3, 40% HF) is added and the vial capped for 15 min. Finally, ethanol (5 cm^3) and 1% butyl-PBD in toluene (5 cm^3) are added and the sample counted as a homogeneous solution.

Scraping material off a TLC plate is, of course, a hazardous operation because of the danger of small radioactive particles being ingested as they fly into the atmosphere. The use of a small 'vacuum-cleaner' type apparatus as shown in Figure 4.12 limits the danger to some small extent, although the collected powder must still be transferred to the scintillation vial.

An alternative approach is to 'fix' the powder before it is removed from the plate. One ingenious way of doing this is to coat the plate with a quick setting film based on cellulose acetate. The film adheres to the solid material of the TLC plate, thus eliminating any possibility of the production of loose powder particles when samples are taken from the plate. The procedure is as follows. Approximately 20 cm^3 of a solution of cellulose acetate (7 g), diethylene glycol (3 g), camphor (2 g), *n*-propanol (25 cm^3) and acetone (75 cm^3) are spread over a 20 cm × 20 cm TLC plate with a glass rod. When the mixture is dry it is cut into squares which then curl up on the surface of the plate, taking the absorbent with them. The squares are then simply transferred into a scintillation vial (with forceps) for assay directly (see next section).

(b) Counting of material on discs or sheets Very often it is convenient to count samples of solutions or suspensions of non-volatile radioactive materials on solid supports which can be introduced into a scintillation mixture (usually small discs made of a variety of materials such as filter paper, glass fibre, cellulose acetate or nitrate – 'Millipore' – filters, expanded plastic, etc.). Normally, samples of the materials are either spotted and dried on to the disc (volumes of between 0.1 and 0.3 cm^3) or samples are filtered through the material

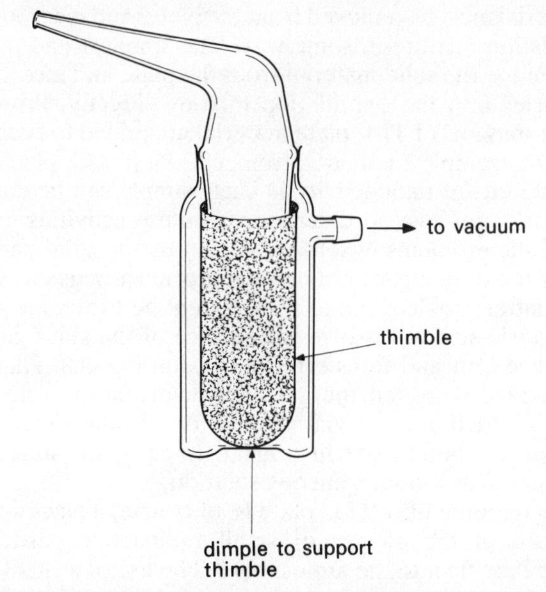

Figure 4.12 Simple vacuum-cleaner-type apparatus for removing radioactive material from TLC plates. The plate substrate is first loosened with a spatula and then sucked into the thimble. Once collected the powder can be carefully transferred to a counting vial or processed further.

and then dried (volumes dependent upon the nature of the sample). The discs can then be assayed simply by covering them with a toluene-based scintillant mixture (e.g. 0.5% butyl-PBD in toluene) and counted. The efficiency of counting on discs varies considerably with the nature of the disc material and sample. Self-absorption is very important and, in particular, for ^{3}H samples is determined by the thickness and absorptivity of the solid support. For example glass fibre discs, cellulose acetate paper and cellulose chromatography paper give ^{3}H counting efficiencies of about 10%, 6% and 2% respectively, depending to some extent on the size of the molecules in the sample. Efficiencies can be increased by solubilizing samples from the discs or dissolving both discs and samples. Various methods are available to achieve this depending on the nature of the sample and disc material and the reader should consult more specialized references as to the methods currently adopted. What is important, however, is to ensure that samples are *either* totally insoluble and are counted as such *or* that they are totally soluble. In the circumstance that a sample is only partially soluble counting errors are introduced depending upon the amount of sample present, and this situation should be avoided. A simple test to ascertain that a sample is insoluble in scintillation fluid is to remove the disc from its original vial and recount the vial. Obviously, if the sample is insoluble, then the recounted vial will only show a background count. An equally simple procedure can be worked out for establishing the complete solubilization of a sample for a disc should this have been attempted. It is not usual for this method of counting on solid supports to be used in situations where the level of radioactivity must be accurately assessed. This is because the level of quenching is difficult to determine since the radioactive material is in the solid phase, and problems such as self-absorption, sample preparation, the nature of the disc material, and orientation of the disc in the vial are all important. However, counting on discs has the advantage that sample preparation is quick, the discs are easy to handle and more than one disc can be introduced into a single counting vial if necessary. For example, using glass fibre discs it is possible to add up to twenty-six discs per vial, which means that at ~0.3 cm^3 sample size per disc a total of 7.5 cm^3 of original sample can be assayed.

(c) Counting polyacrylamide gels Polyacrylamide gels are commonly used for the analysis and characterization of proteins and nucleic acids. Such gels are normally insoluble in liquid scintillation fluids and so special techniques are required for counting them.

Gels can be counted directly if they are reduced to fine particle size before incorporation into the scintillation fluid. This limits the amount of self-absorption of the β-particles but requires special apparatus to achieve fine particle sizes. An alternative method for direct counting is to replace the water in the gel with the liquid scintillant. The gel is cut into slices (about 1 mm thick) and the separated protein or nucleic acid fixed by soaking the slices in acetic acid containing phosphotungstic acid. A further soaking in a mixture of glacial acetic acid and ethylene glycol monomethyl ether (1 : 1) removes the residual water in the gel, which is then placed in scintillator solution. Finally the gel slice is added to fresh scintillator solution in a vial for counting. Each stage of soaking requires 3–5 h or more. Other methods of counting rely either upon completely or partially solubilizing the gel or on complete oxidation. Solubilization is most easily achieved though the use of commercial tissue solubilizers such as NCS or Soluene. Gels cross-linked with bisacrylamide are the most difficult to dissolve and treatment with a tissue solubilizer at a high temperature (60–65 °C) for periods up to 24 h normally results only in a swollen gel slice. However, this treatment releases 90–95% of the counts from the slice and can yield reproducible and stable high efficiency measurements.

Polyacrylamide gels cross-linked with ethylene diacrylate are easier to dissolve and incubation times of 1–5 h at 37°C with the appropriate tissue solubilizers can give complete solubilization, allowing the sample to be counted subsequently in a toluene-based scintillation mixture. *N,N'*-Dialkyltartaramide-linked gels can be completely dissolved in 2% periodic acid (1 cm^3 per 30–50 mg gel) in 20–30 min at room temperature. The resulting clear solution can then be mixed with an emulsion-type scintillator solution and counted.

Combustion of the gel slices represents yet another method by which the slices can be counted. Various 'wet' combustion techniques exist, but all suffer from the danger that radioactive gases may be lost from the solution. Perhaps the best method is to use a sample oxidizer (see p. 66) if this is available. In this technique the gel slice is simply dried down on a small filter paper. The dried gel and filter paper is then burnt in the oxidizer and any radioactive CO_2 and H_2O trapped and counted. The advantages of this method are that double-labelled samples can be counted at high efficiencies with no danger of excessive quenching due to the presence of material from the gel slice or sample.

4.7 Autoradiography

Autoradiography can be defined as a method for locating the position of radioactive substances by the use of modified photographic techniques. The first autoradiographic experiment using radioactive material was unwittingly performed by Henri Becquerel in the 1890s when he was investigating the possibility that fluorescent compounds such as uranium phosphate and uranium sulphate might give off radiations similar to the so-called roentgen rays. Part of this experimental technique consisted of placing the fluorescent minerals on top of a photographic plate which was itself protected from exposure to light by several layers of black paper. After exposure of the mineral-photographic plate 'sandwich' to sunlight for several days, subsequent development of the photographic plate showed dark spots immediately below the position of the minerals. By a happy stroke of fortune Becquerel performed an experiment in which the minerals were juxtaposed to the photographic plate in the absence of sunlight. Subsequent development of the plate still showed dark spots due, as we know today, to the inherent radioactivity in the uranium salts he was using. He had performed the first autoradiographic experiment using radioactive material!

Since that time the technique of autoradiography has become widely used for locating the position of radioactive materials in or on specimens. In biology the technique has become one of the indispensable 'tools of the trade' and is used both at the macro- and microscopic level. There are many excellent monographs detailing the methods used in autoradiography and the reader intending to employ this technique is recommended to consult one of these (see Appendix D). Here we shall just briefly outline the theory of the methods. Some of the techniques currently employed are also discussed on p. 116.

Basically, if a piece of tissue containing radioactivity is placed in close contact with a photographic emulsion for a suitable length of time, radiation from the material will penetrate the emulsion and form a track marking the path of the particle. Following development of the emulsion, providing it remains or is replaced in the same position on the specimen, the areas containing radioactivity in the specimen can be identified. Figure 4.13 shows a typical autoradiogram produced using a bean plant labelled with $^{14}CO_2$ (see also Exp. 4.8.2, p. 74).

It should be noted that the nature of the radiations determines the appearance of the final autoradiograph. In the emulsion only grains receiving enough energy from the radioactive particles will be developed. Thus α-particles produce straight, short, thick lines in a

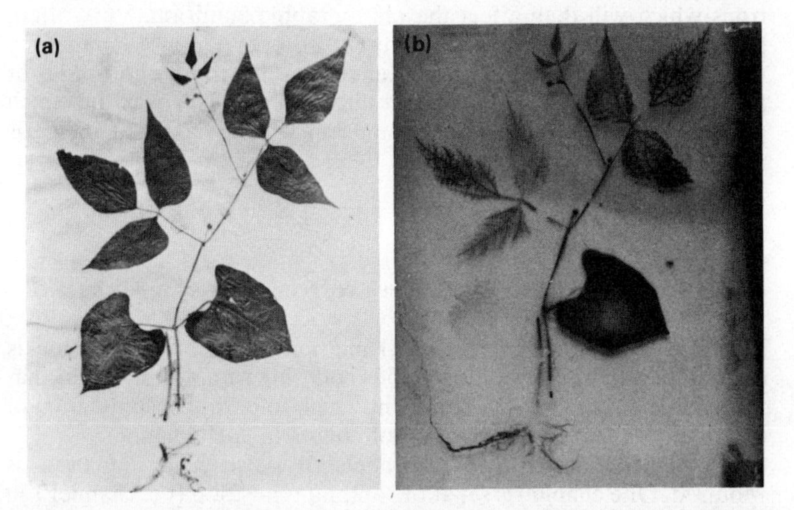

Figure 4.13 Autoradiogram of a 3-week-old bean plant, one leaf of which has been exposed to 370 kBq (10 μCi) of $^{14}CO_2$ in a leaf chamber for 2 h. (a) Dried, pressed plant. (b) Autoradiogram: the fed leaf (FL) appears heavily labelled and radioactive material has clearly been transported to other regions of the plant.

Table 4.5 General properties of particles and radiations used in autoradiography.

Radiation	*Mass (ratio)*	*Energy (relative)*	*Penetrability (relative)*	*Specific ionization (relative)*	*Range in*
α-particle	7380	moderate–high	1	10 000	μm
β-particle	1	low–high	100	100	μm–mn
γ-ray	0	high	100–10 000	1	no limit

photographic emulsion since they are large, heavy and traverse only a short path, but produce large amounts of ionization as they do so. γ-rays interact hardly at all, traversing a long path, producing relatively little ionization and so not activating the photographic emulsion. β-particles are somewhere in between, producing a thin spidery track for radiations from the more energetic isotopes or producing one or two single grains for the weaker ones such as tritium. Table 4.5 indicates the general properties of particles and radiations used in autoradiography. Where it is necessary to use a γ-ray emitting isotope, special techniques are used to generate weak secondary electrons which will then affect the photographic emulsion. Very often, however, it is simpler to use a different isotope, and most commonly the isotopes employed in biological studies are those emitting α- or β-particles which will readily produce tracks. The way in which particles affect a photographic emulsion to form a latent image is shown in Figure 4.14.

4.8 Practical experiments

4.8.1 Counting efficiency – quench correction by sample channels ratio

The sample channels ratio (SCR) method of quench correction is extremely reliable for homogeneous counting samples provided that the sample count rates are sufficiently high to permit accumulation of a statistically acceptable number of counts in each channel.

An instrument with two channels of pulse height analysis is required. One channel is set at the optimum position (i.e. channel 1 in Fig. 4.9, p. 58). Selection of the second channel is arbitrary and can be varied quite widely depending upon the range of quenching to be covered. For the purpose of this experiment a wide range of quenching is to be covered and channel 2 is selected to give approximately one-third of the total counts recorded in channel 1 for the unquenched standard.

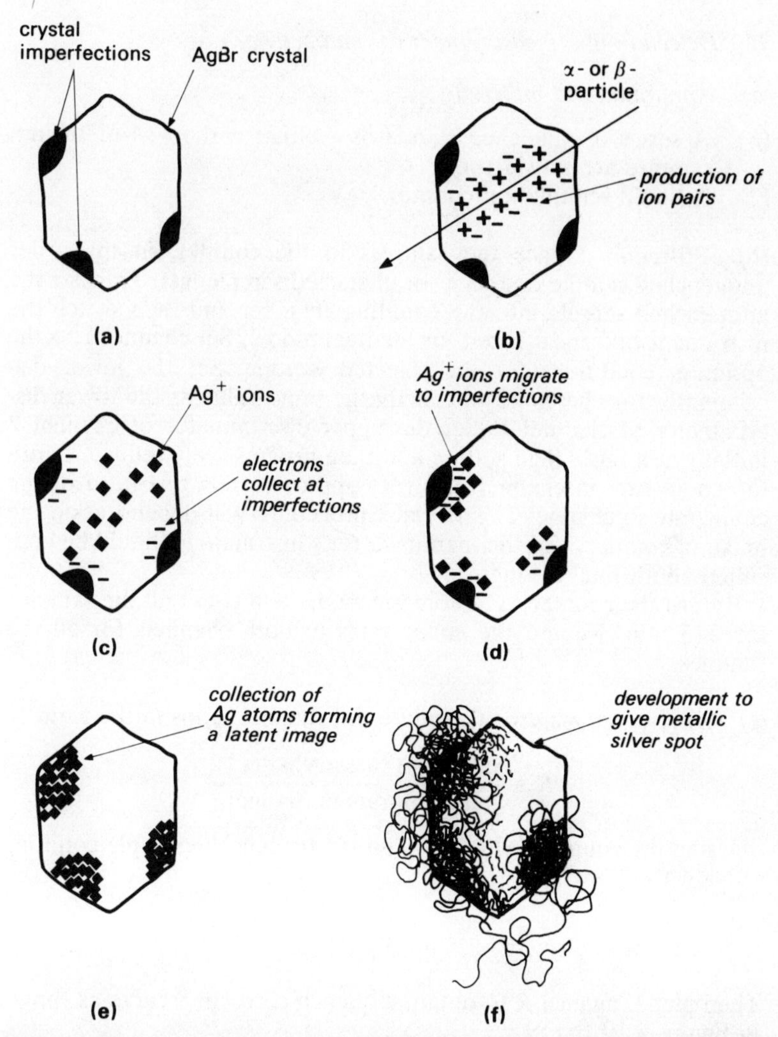

Figure 4.14 A summary of the steps involved in the production of a latent image in autoradiography. Once the latent image is formed subsequent development produces a metallic silver spot at that position. (a) Silver bromide crystals in the photographic emulsion contain 'imperfections' consisting of minute quantities of impurities. (b) The passage of a radioactive particle through the crystal produces free electrons and positively charged silver ions in the crystal lattice. (c) The free electrons tend to collect at the imperfections. (d) Subsequently the mobile positively charged silver ions migrate to these negatively charged regions. (e) Neutralization of the silver ions in these regions produces metallic silver which can act as loci for further silver production. (f) Development of the 'latent images' in the emulsion produces further silver deposition and a visible darkening in the emulsion which represents the initial passage of the radioactive particle.

(a) Apparatus and materials

(1) A series of quenched standards – either carbon-14 or tritium samples are satisfactory
(2) A liquid scintillation counter.

(b) Procedure Place the samples in the counter, in the order unquenched sample first to most quenched sample last. Transfer the unquenched sample into the counting chamber and then switch the instrument to 'repeat count' or 'manual mode'. Set channel 1 to the optimum conditions for the selected isotope. Set the lower discriminator for channel 2 to exactly the same value as the lower discriminator of channel 1. Set the upper discriminator of channel 2 initially at a fairly high setting and then progressively reduce it until the count rate in channel 2 is very approximately one-third of the count rate in channel 1. The exact procedure will depend upon the make of counter, and the manufacturer's instructions should be consulted about final settings.

Return the counter to 'automatic mode' and count all the samples for 2–3 min. Record the count rates in both channels for all the samples.

(c) Analysis of results Calculate the channel ratios in the form

$$R = \frac{\text{net count rate in channel 2}}{\text{net count rate in channel 1}}$$

Using the counts in channel 1 only, calculate the sample counting efficiencies:

$$E = \frac{\text{net sample count rate}}{\text{absolute activity}}$$

Then plot E against R to obtain a quench correction curve as shown in Figure 4.10 (p. 59).

The activities of 'unknown' samples can now be determined by using the calibration curve so obtained.

4.8.2 Autoradiographic detection of photosynthesis and translocation of photosynthate

(a) Apparatus and materials

(1) Leaf chamber (see Fig. 4.15)
(2) Lactic acid (50%)
(3) 1 cm^3 hypodermic syringe

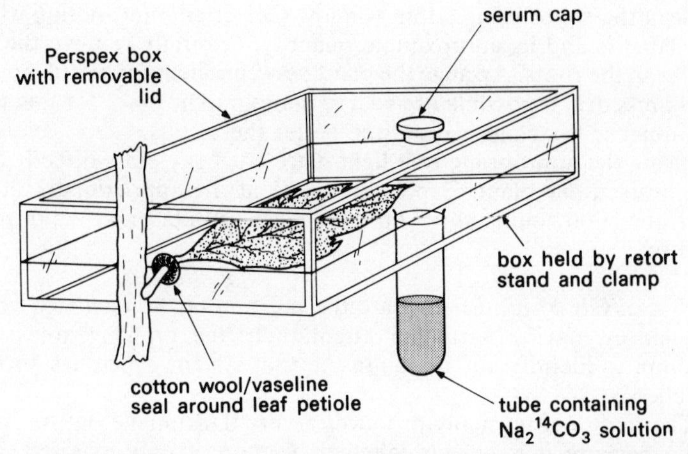

Figure 4.15 Diagram of leaf chamber used for feeding radioactive $^{14}CO_2$ to a bean leaf. The leaf is easily enclosed in the chamber by bringing together the two halves of the apparatus with the leaf in between so that the leaf petiole fits in the semi-circular notches cut into each chamber side. Any gaps between the petiole and chamber are sealed with a plug of vaseline and cotton wool.

(4) X-ray plates
(5) Large sheets of blotting paper
(6) 370 kBq $Na_2\,{}^{14}CO_3$ (10 µCi)
(7) Two 100 W lamps
(8) X-ray developing and fixing solution
(9) Dwarf bean plant (2–3 weeks old).

(b) Procedure Place a few millilitres of water in the leaf chamber (to prevent excessive transpiration) and insert a suitable leaf into it (see Fig. 4.15). Grease all joints with vaseline. Remove the serum cap from the apparatus and place the radioactive sodium carbonate solution (total of 370 kBq) into the tube. Replace the serum cap and inject 0.5 cm^3 of lactic acid into the tube through the cap. Do not withdraw the needle immediately but gently warm the acidified carbonate with a lighted match and using the syringe, 'pump' the generated CO_2 into the chamber. Illuminate the leaf (for several hours) using the two 100 W bulbs, after which time the leaf chamber should be carefully dismantled, preferably in the open air or in a fume cupboard (5% KOH solution can be injected into the chamber before dismantling to remove any residual $^{14}CO_2$).

Scan the shoot using a thin window Geiger counter, noting where the label is and its approximate quantity. Carefully remove the soil and scan the roots. Arrange the plant between sheets of blotting paper and press dry. If possible use a large photographic dryer for this since the quicker the plant is dried the better the results.

Using the appropriate safe light place an X-ray plate directly over the pressed *dry* plant. Expose the plate at the approximate rate of 24 h for 1000 counts min^{-1} on the Geiger counter and then develop and fix.

(c) Analysis of results Locate the position of radioactivity within the leaves, paying particular attention to the growing areas, and attempt to identify any region in the root system which has become labelled.

The 'fed' leaf normally produces a very dark image on the X-ray plate because it is heavily labelled. A more evenly exposed autoradiogram can be produced, if required, by interposing a thin sheet of cellophane between the 'fed' leaf and the X-ray plate during exposure.

5 Factors affecting the design of tracer experiments

5.1 Errors, statistics and data reduction

The results of radiotracer experiments are initially recorded as a count rate (counts per unit time). Count rates are arbitrary units and values depend upon the type of instrument being used to record the radioactive distintegrations. For results to be meaningful they must either be related to a standard (most tracer work employs this method) or converted to a value representing the absolute amount of radioactivity present.

It is necessary, therefore, to apply corrections to measured readings. Such corrections include: (a) an assessment of the number of disintegrations in the sample which have not been recorded, (b) an allowance for counts arising from background radiation separate from the sample, and (c) losses due to radioactive decay (mainly for short-lived isotopes).

5.1.1 Corrections for counts not recorded

(a) Corrections for resolving time When radiation interacts with a detector there is a short period (typically ~100 μs for a Geiger–Müller counter) during which no further radiation can be detected. This is known as the resolving time of the apparatus. Normally counting systems are designed with a fixed resolving or 'dead' time which can then have a precisely defined value rather than leaving it to be determined experimentally. This means that the magnitude of the loss due to dead time can be calculated.

For example, suppose we have a sample which has an observed count rate of 200 counts s^{-1} using an instrument which has a dead time of 100 μs. The counter 'switches off' for 100 μs after every count is recorded, and therefore while recording the 200 counts it has been off for a total of 200 × 100 μs (= 0.02 s). The true counting rate is therefore

$$\frac{200}{1 - 0.02} = \frac{200}{0.98} = 20\dot{4} \text{ counts s}^{-1}$$

More formally, we can show that if C is the true count rate, c the observed count rate and t the resolving time then

$$C = \frac{c}{1 - ct} \tag{5.1}$$

Resolving time is only significant if it is long and count rates are high in relation to it. In tracer work it is usually only important when using GM counting with highly active samples ($C > 5000$ counts min^{-1}).

(b) Correction for efficiency of counting Even allowing for resolving time, detectors cannot be 100% efficient and will therefore not pick up all of the disintegrations in the sample.

The special methods for determining counting efficiencies are described in Chapter 4 in conjunction with the appropriate type of counter to which the method applies, but it should be noted here that a wide range of factors affect counting efficiency. For GM work these include back-scatter, self-absorption in the source, absorption in the counter window and absorption in the air gap between source and detector; for liquid scintillation counting efficiency is greatly affected by 'quenching' and even very similar samples will have slightly different counting efficiencies. However, with liquid scintillation counting, standardization is fairly straightforward.

For example, suppose we obtain an uncorrected count of 4500 counts min^{-1} using a liquid scintillation counter whose background is 30 counts min^{-1}. Suppose also that reference to a channels ratio quench correction curve gives an efficiency $E = 0.65$, then the absolute activity in the sample (with respect to the standard) is given by

$$\frac{4500 - 30}{0.65} = 6877 \text{ disintegrations min}^{-1}$$

With end window GM counting and crystal scintillation counting determination of absolute efficiencies is very difficult and hence the number of disintegrations per minute are not obtained. It is most usual to 'standardize' count rates with respect to a standard source as described in Chapter 4.

Suppose a standard source gave a net count rate of 5600 counts min^{-1} when an experiment first started. Some weeks later a sample was counted and gave 2300 counts min^{-1}. Immediately following the sample the standard was counted and now gave 5480 counts min^{-1}, which indicated a slight change in counting efficiency during the period of the experiment. The latter value is used to correct the sample count rate to give a 'standardized count rate' of $2300 \times 5600/5480 = 2350$ counts min^{-1}.

5.1.2 Corrections for background radiation

Some recorded counts will arise, not from the sample, but from cosmic and other background radiation and also from electronic noise within the counter itself. These counts must be subtracted from the observed total count to give the sample count itself. When sample count rates are high and background is low the correction is of little importance, but in cases where sample counts are low (very often the case with biological material) then background correction is essential.

When very low activity sources are being measured, very long counting times may be required to achieve an acceptable statistical accuracy (see counting statistics, p. 80). In such cases it is best to choose the most efficient distribution of time between counting background and background + sample counts. The highest accuracy is achieved when

$$\frac{t_1}{t_2} = \sqrt{\frac{(C + B)}{B}} \tag{5.2}$$

where t_1 is time spent counting sample + background, t_2 is time spent counting background, $C + B$ is the sample + background count rate, and B is the background count rate.

Thus, for example, if the (sample + background) rate was 120 counts min^{-1} and the background rate was 30 counts min^{-1}, then the ratio of (sample + background) time to background counting time should be

$$\sqrt{\frac{120}{30}} = \sqrt{4} = 2$$

i.e.

$$t_1 = 2t_2.$$

5.1.3 Correction for radioactive decay during the experiment

These corrections obviously only apply to experiments using short lived isotopes or to experiments which are conducted over long periods of time. In biological work the most commonly used short lived isotopes are ^{35}S, ^{125}I and ^{32}P, and here it is very often necessary to apply corrections to experiments in which they have been employed. Such corrections may be made using published decay tables or by direct comparison of sample count rates with a standard containing the same isotope and which is counted at the same time as the samples. Alternatively the calculation can be carried out using the decay equations derived in Chapter 2 (p. 18). Consider an experiment with ^{131}I ($t_{\frac{1}{2}}$ = 8.04 d) in which a sample gives 2800 counts min^{-1}, 5 days and 2 h (122 h) after the initial count was recorded. Rearranging equation 2.5 gives

$$\lambda = \frac{\ln 2}{8.04 \times 24} = \frac{0.693}{192.96} = 3.59 \times 10^{-3}\ \mathrm{h}^{-1}$$

Then taking equation 2.4 and writing a_0 for n_0 and a_t for n_t (since from equation 2.3 $a = dn/dt$ and therefore from equation 2.2 a is proportional to n)

$$\ln \frac{a_t}{a_0} = -\lambda t = -3.59 \times 10^{-3} \times 122$$
$$= -0.4380$$

therefore

$$\frac{a_t}{a_0} = 0.645$$

(this would be the figure obtained from a decay table); therefore

$$a_0 = \frac{2800}{0.645} = 4341\ \text{counts min}^{-1}.$$

5.1.4 Counting statistics

(a) General considerations The decay of radioactive material is a random process. This means that the number of disintegrations observed in any time period is not fixed but will vary over a range, the extent of which depends upon the length of the time of measurement. The values of count rates obtained by successive counts of a long lived isotope correspond to a Poisson distribution; for most practical purposes these approximate to a normal or Gaussian distribution (Fig. 5.1). In Figure 5.1 σ represents the standard deviation and $\bar{c}$ the mean count rate. The mean count rate is given by the formula

$$\bar{c} = \sum_{i=1}^{N} \frac{C_i}{N}$$

(where N is larger than $\bar{c}$, the arithmetic average of N counts approaches the true mean, although the 'true' value is never so

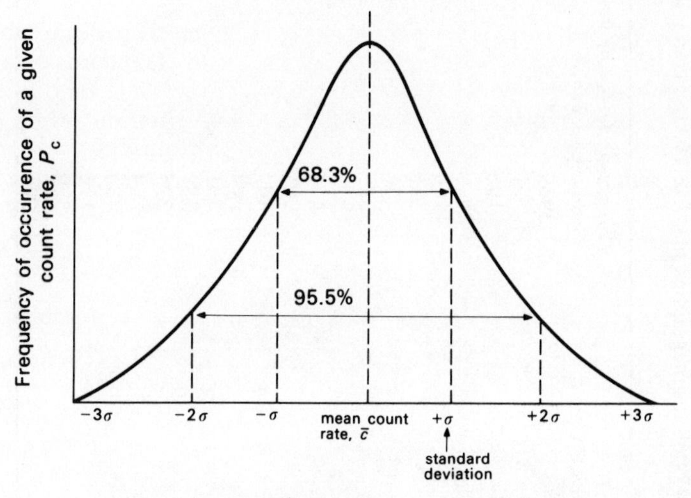

Figure 5.1 A plot showing the normal or Gaussian distribution of count rates. Count rates actually follow a Poisson distribution but this approximates to a normal or Gaussian distribution for most practical purposes. (If the measurements in Table 5.1 were to be repeated many times and the results plotted as frequency of occurrence of a given count rate against count rate, then a similar plot would be obtained.)

obtained). For example, if we apply this formula to the data in Table 5.1 we have

$$\bar{c} = \frac{136\,316}{20} = 6816 \text{ counts in } 0.5 \text{ min}$$

i.e. for this data the average count rate $\bar{c}$ is $6812 \times 2 = 13\,632$ counts min^{-1}.

The standard deviation gives information about the width of the distribution function. Of the measured values of activity, 68.3% lie within $\pm\sigma$ of the mean value $\bar{c}$, 95.5% within $\pm 2\sigma$ of the mean and 99.7% within $\pm 3\sigma$ of the mean.

For a set of N measurements

$$\sigma = \left[\frac{1}{N-1} \sum_{i=1}^{N} (C_i - \bar{c})^2\right]^{\frac{1}{2}}$$

Appplying this formula to the data of Table 5.1 we have

$$\sigma = \left[\frac{1}{19} \times 14\,218\right]^{\frac{1}{2}} = 27$$

Table 5.1 Sequence of counts from a tritium sample having an activity of 47 068 disintegrations $min^{-1} \pm 1\%$ and counted at approximately 30% efficiency.

Measurement no., i	*No. counts in 0.5 min interval, C_i*	*Deviation of counts from mean, $C_i - \overline{c}$*	*Square of deviation from mean, $(C_i - \overline{c})^2$*
1	6879	63	3969
2	6815	− 1	1
3	6830	14	196
4	6817	1	–
5	6782	−34	1156
6	6800	−16	256
7	6822	6	36
8	6833	17	289
9	6816	0	0
10	6842	26	676
11	6785	−31	961
12	6796	−20	400
13	6853	37	1369
14	6785	−31	961
15	6799	−17	289
16	6801	−15	225
17	6818	2	4
18	6783	−28	784
19	6862	46	2116
20	6793	−23	529
Sum	13 6316		14 218

i.e. for this data the standard deviation is ±27. In other words, if we were to make one more measurement on the sample used for the results in Table 5.1, then in 68% of the observations the results would be in the range 13 632 ± 27 counts min^{-1} ($\bar{c} \pm \sigma$ counts min^{-1}) and in 95% of the observations the result would be in the range 13 632 ± 54 counts min^{-1} ($\bar{c} \pm 2\sigma$ counts min^{-1}). The standard deviation is thus a measure of the precision of the observations.

(b) Count rates for single measurements Most often in isotope work only one measurement on a single sample is made. How then is the standard deviation of the distribution estimated? If we record c counts then we can say that this represents our best estimate of $\bar{c}$ and the standard deviation on the count may be taken as $\sqrt{c}$. Normally we would actually require an estimate of the count rate. If c counts had been recorded in time t then the count rate would be

$$\frac{c}{t} \pm \frac{\sqrt{c}}{t} \text{ (with 68\% confidence)}$$

For example, using the average value of the count rate in Table 5.1, then the mean count rate would be 13 632 ± 165 counts min^{-1}.

For single samples greater precision can only be achieved by increasing the total count recorded. For example, suppose we record a reading of 100 counts in 1 min for a sample,

$$\begin{aligned}\text{Counting rate} &= \frac{100}{1} \pm \frac{\sqrt{100}}{1}\\ &= 100 \pm 10 \text{ counts min}^{-1}\end{aligned}$$

i.e. a precision of ±10%.

If, however, we were to accumulate a total of 10 000 counts, say, in 100 min, then

$$\begin{aligned}\text{Counting rate} &= \frac{10\,000}{100} \pm \frac{\sqrt{10\,000}}{100}\\ &= 100 \pm 1 \text{ counts min}^{-1}\end{aligned}$$

i.e. a precision of ±1%.

Thus, although the count rate is the same, the precision of measuring it has been improved. Where possible a good 'rule of thumb' for counting is to ensure that the counting period is long enough to give at least 10 000 counts which thus gives a reading to ±1% with 68% confidence or to ±2% with 95% confidence. For low activity samples, where counting times would be too long for this to be a reasonable requirement, then a lower precision must be accepted.

The above considerations have taken no account of background activity. If we obtain c counts in time t_1 due to sample and background, and background alone gives b counts in time t_2, then

$$\text{count rate} = \frac{c}{t_1} - \frac{b}{t_2} \pm \sigma_b$$

where

$$\sigma_b = \sqrt{\frac{c}{t_1^2} + \frac{b}{t_2^2}}$$

For example, suppose a sample gave 8000 counts in 4 min using a GM counter and a background count with the same machine gave 300 counts in 10 min, then

$$\text{corrected count rate} = \frac{8000}{4} - \frac{300}{10} = 2000 - 30$$

$$= 1970 \text{ counts min}^{-1}$$

and

$$\text{standard deviation} = \sqrt{\frac{8000}{16} + \frac{300}{100}}$$

$$= \sqrt{500 + 3}$$

$$= \pm 22.4 \text{ counts min}^{-1}$$

i.e. count rate = 1970 ± 22.4 counts min^{-1}.

(c) Disregarding abnormal data The random nature of radioactive decay means that there is bound to be uncertainty in the experimental data. Indeed, if we are quoting values with a 68.3% confidence limit (i.e. $\pm\sigma$), then for 32% of the time our measured value will be outside this range! In addition to this, other factors such as variation in sample preparation and measuring errors can contribute to large variations in measurements. Occasionally then we may find that one measurement differs from the others by a large amount which can introduce a large error into an overall average if only a few results are being considered. If data is suspect a reasonable criterion to use is to

reject values which differ by more than 2σ or 3σ from the mean. It should be noted, however, that if more than one value is suspect then the mean and σ values must be recalculated after each value has been rejected and additionally that if a large number of values can be discarded in this way then it must be suspected that the distribution of values is not normal and the method can no longer be applied.

5.2 Choice of isotope

It is sometimes thought that the utilization of a radioisotope for a particular experiment will produce results which are easy to obtain and easy to interpret. This approach is almost certain to result in disappointment for the experimenter. Radioisotopes should only be used when there is no other relatively straightforward non-radioactive method which will give comparable results. Quite apart from any potential health hazard involved, the use of a radioisotope will necessitate a considerable degree of preparation and forethought concerning the experiment. Also, because biological systems are very often complex and multi-compartmental, full consideration must be given to the kinetic aspects of the system when interpreting results. Despite these limitations it must be said that the use of radioisotopes very often permits elegant solutions to problems which would be difficult to solve in any other way.

Having decided that a radioisotope is necessary, how does one decide which one to use? For most biological applications it is fairly obvious which element is the one of choice. For example, if we wished to follow the fate of acetate ions in a system then radiolabelled carbon (^{14}C) would be the isotope of choice. Likewise, if we wished to investigate the incorporation of nucleotides into a nucleic acid, we would probably choose precursors radiolabelled with phosphorus (^{32}P).

It is not always necessary to be limited to isotopes of the same elements as those contained in the compound of interest. For example, proteins are very often labelled with iodine ^{131}I or ^{125}I for various types of tracer experiment. When choosing which isotope to use some consideration must be given to its inherent properties such as its half-life and the type of radioactive emission it produces. Isotopes such as ^{14}C and ^{3}H usually present no problem with regard to their half-lives (5730a and 12.26a respectively), but ^{3}H could present a problem regarding its detection if the appropriate equipment were not available. (The weak β-emissions from ^{3}H cannot be detected using an end window GM tube.)

Having chosen which isotope to use the next step concerns the choice of compound in which it is to reside (where such a choice exists). Several considerations apply here. Firstly, one must consider the position of the label within the compound. For tracer studies of metabolic pathways it is important that the positions of the radiolabel within the compound should be known so that its movement through the various metabolic products can be clearly understood. (Where kinetic studies are involved any possible isotope effect must also be accounted for – see p. 94). Where general incorporation is being considered then very often a uniformly labelled or a generally labelled compound will be quite satisfactory. Secondly, the maximum

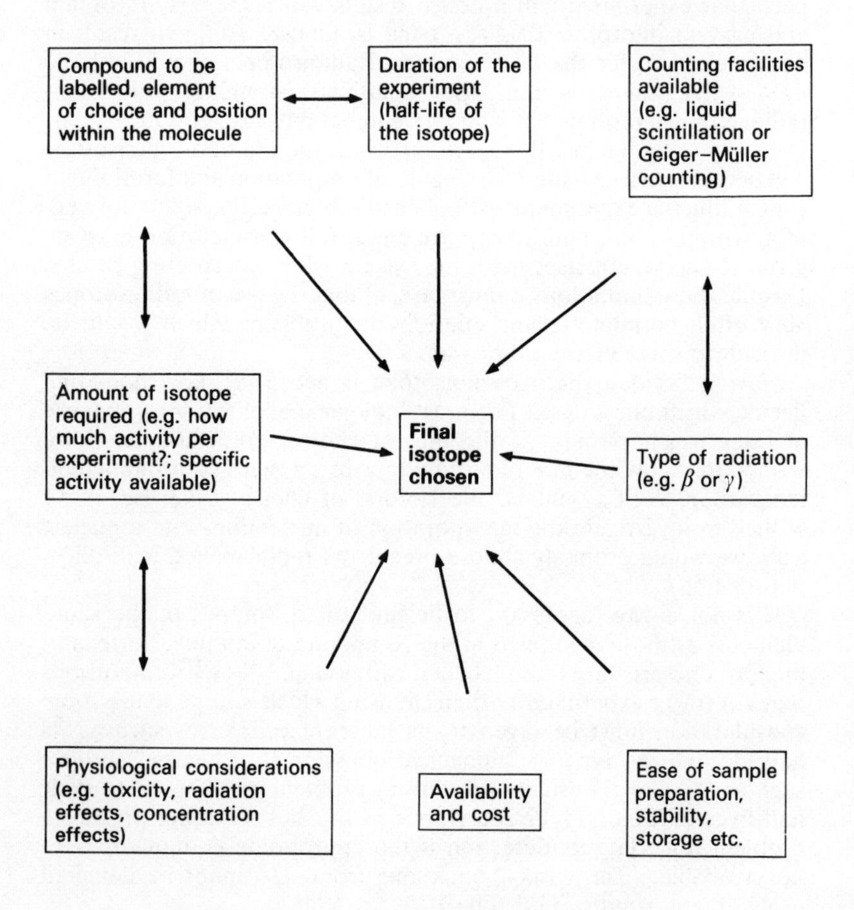

Figure 5.2 Factors influencing choice of isotope.

specific activity available must be considered. If the compound being used exists naturally at a low endogenous concentration, then it is important to use as high a specific activity as possible in order to avoid any non-physiological increase in concentration which may be produced using a compound with low specific activity. The specific activity of the compound must then be considered from the point of view of ease of detection once the reaction has been completed (see amount of activity required, below. Thirdly, and not unimportantly, one must consider the cost of a particular compound. It is no use embarking upon an experiment requiring several megabequerels of isotope which costs £100 kBq^{-1} if sufficient money is not available and if the results are not going to justify the expense of the operation.

Finally, various considerations have to be given to the biological aspects of the system such as the possibility of radiation damage – for example is the system being physiologically affected by the emitted radiation? In this connection it is perhaps worth noting that in some growing systems isotope concentrations as low as 60 $\mu Ci/cm^3$ of ^3H or ^{32}P can stop growth completely. In addition, any chemical addition to the system must not exceed the normal physiological concentrations and amounts and the physical and chemical state of the radioisotope should be identical to the unlabelled compound which it is replacing.

A summary of the various considerations is shown in Figure 5.2.

5.3 Amount of activity required

One problem which frequently confronts the novice in radioisotope work concerns the amount of activity which should be used for any particular experiment. At the outset it should be recognized that it is almost impossible to calculate the exact amount of activity required unless the details and yields for each step of the experiment are precisely known. This is very often not the case and usually one has to resort to a rough 'rule of thumb' calculation. The beginner in radioisotope work should not, however, be put off from performing these calculations since they can indicate the correct order of activity one should use and, in the more extreme cases, show whether an experiment is at all feasible. A typical example might serve to indicate the type of approach used.

Suppose we wish to perform a series of extractions of a growth hormone from an organism in order to determine its amount and concentration. We would need to keep a check on the efficiency of the extraction process since any changes in this would obviously change our estimate of the amount of hormone present. One way to do this would be to add a known amount of radiolabelled hormone at the beginning of the extraction process and subsequently determine its concentration at the end, hence enabling an estimate of the efficiency of the extraction to be measured. How much activity do we need? Suppose we wish to be able to determine the concentration to within ±1%. This requires that we must accumulate at least 10 000 counts from the radioisotope disintegration (see Section 5.1, p. 83), and that the background counts accumulated during the same time period are relatively small. In addition, let us suppose that we are using a hormone which has been labelled with tritium and that we are using a scintillation counter for measurement. A reasonable activity would be one that allowed us to measure 10 000 counts in 10 min (during this time the background measurement would amount to approximately 350 counts). A scintillation counter will be approximately 35% efficient at measuring tritium, so the actual activity will be approximately 30 000 disintegrations in 10 min, or 3000 disintegrations min^{-1}. The sample we finally measure must therefore have an activity of at least this value. Now let us consider the extraction process itself. It is difficult to be precise here (after all, it is this precision that we are trying to measure!), but suppose that the process contains three steps of equal complexity and that each step is 50% efficient. This means that the overall efficiency of extraction is approximately 12%. In addition, suppose that we wish to split our final extraction into three samples for counting. Each sample must contain at least 3000 disintegrations min^{-1}, which means that the finally extracted sample must contain a total of 9000 disintegrations min^{-1}. Allowing for a 12% efficiency of extraction, then the required addition before extraction amounts to approximately 75 000 disintegrations min^{-1} or 1.258 kBq (0.034 μCi). In practice, since a large number of assumptions have been made, it would be better to increase the figure by approximately 50%, i.e. to use 1.85 kBq (0.05 μCi) initially. Although the preceding calculations may appear rough and ready, the approximate value so obtained is of some use for the first experiment; in any case the amount used will probably need to be changed once the actual experiment has been performed and a more accurate idea of the efficiencies involved can be obtained.

In addition to the above we also require to form some idea of the specific activity we need to use. In this case the amount of hormone we add should not contribute substantially to that present in our extract otherwise inaccuracies in estimation might result. Suppose we expect to obtain approximately 10 μg of hormone per extract and that the hormone has a molecular weight of 250 daltons. We can probably add up to 1% of the radiolabelled hormone without any adverse effect on accuracy, i.e. we can add up to 0.1 μg of hormone to our extract. This must contain 1.85 kBq (0.05 μCi). In other words the specific activity must be equal to or greater than 18.5 kBq μg^{-1}

(0.5 μCi μg^{-1}) or 4625 MBq $mmol^{-1}$ (125 mCi $mmol^{-1}$). The calculations we have made may be summarized in the following way:

Restriction	*Decision*
Accuracy required for counting (65% confidence), ±1%	10 000 counts
Time allowed for counting samples, 10 min	1000 counts min^{-1}
Efficiency of counter for ^3H, 35%	3000 disintegrations min^{-1}
No. of samples from extract, 3	9000 disintegrations min^{-1}
Efficiency of extraction process, 12%	75 000 disintegrations min^{-1} ≡1.25 kBq (0.034 μCi)
Allowance for 50% error in estimation	1.85 kBq (0.05 μCi)
Amount of substance we can add, 0.1 μg	specific activity = 18.5 kBq μg^{-1} (0.5 μCi μg^{-1})
Molecular weight of substance, 250 daltons	specific activity = 4625 MBq $mmol^{-1}$ (125 mCi $mmol^{-1}$)

A general summary of the decisions which need to be made when choosing an isotope for a particular experiment is given in Table 5.2.

Table 5.2 Summary of decisions to be taken in choosing the amount of isotope to use. C and D together give an estimate of the minimum specific activity required.

Factors	Outcome	
(1) Choice of isotope	dictated by nature of experiments and experimental material	A
(2) Accuracy required (3) Efficiency of counting (4) Acceptable counting time (5) Sample size	give requirements of counts per sample	B
(6) Dilution of labelled compound by experimental system (7) Losses in preparation of sample (8) Corrections for decay of isotope	give requirement of number of counts to be added to the system	C
(9) Endogenous level of compound if occurring naturally (10) Physiological considerations – toxicity, permeability, etc.	give amount of compound which can be added to the system	D

5.4 Stability and storage of radioactive compounds

When the results of an experiment which has utilized a radiochemical tracer are considered it is very important to take account of the radiochemical and radionuclide purity of the starting material. Absolute purity is something which cannot be achieved in practice, but it is essential that the tracer used does not contain even small amounts of contaminating substances which will specifically interfere with the reaction under investigation either through some chemical reaction or through the incorporation of an unwanted radiochemical species. For example, it has been shown that the self-decomposition products of [^{3}H]thymidine (which is often used for studies on DNA synthesis) can be incorporated into macromolecules other than DNA. Thus, autoradiography of *Tetrahymena pyriformis* cells using impure materials showed extensive labelling of the cytoplasm, whereas only labelling of the nucleus (nuclear DNA) with perhaps slight labelling in the cytoplasm (in mitochondrial DNA) would have been expected using pure [^{3}H]thymidine.

One of the factors which contributes to the level of impurity in a given radiochemical is the mode and length of time of storage, since decomposition of the compound during storage will inevitably introduce impurities.

There are four ways in which the decomposition of a radiochemical may take place. Firstly, and unavoidably, there is the decomposition arising from the disintegration of the unstable nucleus of the radioactive atom (called primary (internal) decomposition). For singly labelled molecules decay to the stable isotope will result in a very low concentration of a non-radioactive compound. For multiple-labelled molecules decay will produce new radioactive compounds, but in general the amounts will be extremely small. For example, only 0.001% per year of [^{14}C]methylamine is formed from [1,2-^{14}C]ethane as a result of normal radioactive decay:

$$^{14}CH_3 - {}^{14}CH_3 \rightarrow [{}^{14}CH_3 - {}^{14}NH_3] \rightarrow {}^{14}CH_3 - NH_2$$

The second mode of decomposition is called primary (external) decomposition. This involves the direct interaction of emitted radiation with other molecules of the radiochemical close to the decaying atom. When the emitted particles strike a labelled molecule the interactions produce radioactive impurities, otherwise with unlabelled molecules chemical impurities can be formed. Normally this mode of decomposition becomes increasingly evident as the molar specific activity of the compound is increased and as the relative separation of the reacting molecules is reduced. In addition the weaker the energy of the radiation the more it can be absorbed by the compound and the greater the self-decomposition. Thus, in general, for a given specific activity and under the same conditions of storage, molecules containing a tritium label will show more self-decomposition than ones containing a carbon-14 label, which in turn will show more than those containing a phosphorus-32 label.

Perhaps the most damaging mode of decomposition is secondary decomposition. Here labelled molecules interact with free radicals and other excited species produced by the radiation. In aqueous solution the hydroxyl radical seems to be the most harmful species. It can produce, for example, tyrosine from [^{14}C]phenylalanine and glycols from labelled pyrimidines. In organic solvents the mechanisms are likely to be very complex, but the presence of even trace amounts of impurities in the solvent can cause an increase in self-decomposition. Particularly important are peroxides, small amounts of which may destroy the labelled compound completely.

Finally, we must consider chemical and microbiological decomposition. Radiochemicals are often stored and used in solution at very low chemical concentrations. Such low concentrations render them susceptible to decomposition in ways which would not normally be considered as important. For example, even glass containers can affect the stability of a radiochemical through interaction with the glass surface, however well cleaned. Many labelled compounds are excellent substrates for microbiological growth and any contamination by a microorganism could very quickly affect the rate of decomposition of a radiochemical in solution, even if it did not completely destroy the labelled compounds.

Since there are so many possible ways in which a radioisotope may be subject to unwanted decomposition it is obvious that some control over decomposition must be maintained during storage. Various methods may be adopted, of which the following are important:

(a) Reduction of the molar specific activity of the compound by diluting it with unlabelled molecules of the same compound. This is obviously only of use when low specific activities can be used.

(b) Dilution of the compound in a suitable solvent. This stabilizes the compound by lowering the radioactive concentration and has the additional advantage that it is often convenient to dispense a radiochemical from its solution in some suitable solvent. Very often manufacturers supply a radiochemical in an appropriate solvent which is compatible with the compound concerned and does not cause problems with regard to free radical formation.

(c) Free radical scavengers are sometimes included in a solution

Table 5.3 Recommended storage conditions for labelled compounds.

Isotope	*Compound*	*Specific activity*	*Dispersion*†	*Temperature (°C)*	*Maximum concentration (MBq cm^{-3})*
tritium‡	amino acids	low	FD solid	+2	—
	amino acids	high	AS: can contain 2% ethanol	+2	37
	carbohydrates and nucleosides	low and high	AS: can contain up to 10% ethanol	+2	37
	nucleotides	low and high	50% aqueous ethanol	−20	37
	steroids	—	benzene containing ethanol	+2 or +20 (unstable ones at −140 to −196)	37
	catecholamines	—	AS	−140 to −196	<37
	others	—	AS, benzene or ethanol	+2 or +20	—
phosphorus-32	nucleotides	low	FD solid	−30	—
	nucleotides	high	FD solid	−140	—
	nucleotides	high	50% aqueous ethanol	−20 to −40	186
sulphur-35	amino acids	high	AS containing 5 mM 2-mercaptoethanol (under nitrogen)	−140	185
	others	—	natural form	+20 or lower	—

carbon-14	amino acids	low	FD solid	0	—
		high	AS containing 2% ethanol	−20 or 0	1.85–3.7
	carbohydrates and nucleosides	low	FD solid or syrup	0	—
		high	AS containing 2–3% ethanol	−20	1.85–3.7
	nucleotides	low and high	AS containing 2% ethanol: pH 7	−20	1.85–3.7
	steroids	low and high	benzene solution containing 5 or 10% ethanol	+2 or +20 (unstable ones at −140 to −196)	—
	catecholamines	—	FD solid or AS	−20 (unstable ones at −140 to −196)	—
	others	—	AS or ethanol	0	—
			benzene	+2 or +20	—

†AS, aqueous solution; FD, freeze dried.
‡Solutions of tritium compounds should not generally be stored in the frozen condition.

to minimize decomposition. Good free radical scavengers include ethanol, sodium formate, glycerol and ascorbic acid. Obviously the scavenger used must not react with the compound it is protecting and also should have no effect on the system in which the radiochemical is being used as a tracer.

(d) In general self-decomposition of a radiochemical can be minimized by storage at low temperature. Low temperatures decrease the rate of most reactions and hence increase the chemical stability of a radiochemical. Where a solution is frozen it is important to avoid any clustering of solute molecules as freezing takes place (e.g. crystallization) since this would effectively offset the advantages of dilution and lead to an acceleration of decomposition.

(e) Finally, the radiochemical should be stored in the correct environment for optimum chemical stability, for example at an appropriate pH or, for an oxygen-sensitive compound, under an atmosphere of nitrogen, etc.

A summary of recommended storage conditions of some commonly used compounds is shown in Table 5.3 (data taken from *Amersham Reviews* Nos 7 and 16, see Appendix D).

Although the decomposition of stored radiochemicals can be minimized, the number of ways in which they may be subject to decomposition and thus accumulate impurities is obviously a formidable one. From the practical point of view it is essential that the following steps are taken:

(a) Where possible ascertain the highest acceptable level of impurity for the particular experimental procedure you are pursuing.

(b) Consult the radiochemical supplier's data concerning the purity of the compound.

(c) Where necessary, and particularly when the compound has been stored for any length of time, purify the compound immediately before use. (Very often the supplier's data sheet gives information on purification – for most compounds a simple chromatographic procedure is all that is required.)

(d) When the compound has to be stored, store using recommended conditions where possible.

5.5 Isotope effects

In Chapter 2 (p. 12) it was observed that, chemically, all the isotopes of an element are very similar due to their identical electronic configuration. However, certain properties depend upon mass. Thus, for the diffusion of gases, Graham's law states that the rate of diffusion is proportional to (molecular weight)$^{-\frac{1}{2}}$ and therefore a gas containing a light isotope will diffuse through a porous membrane faster than one containing a heavy isotope. This has been used to effect separation of isotopes; for example using uranium hexafluoride the separation factor for a single diffusion cell is given by

$$\alpha = \left[\frac{\text{mol. wt } ^{238}\text{UF}_6}{\text{mol. wt } ^{235}\text{UF}_6}\right]^{\frac{1}{2}} = \left[\frac{352}{349}\right]^{\frac{1}{2}} = 1.0043$$

and by using many thousands of such cells arranged in a cascade the separate isotopes may be effectively isolated.

Isotopic substitution can also affect the position of equilibrium in a reversible chemical reaction. For a deuterium (D) substituted hydrogen sulphide reacting with water,

$$\text{HDS (gas)} + \text{H}_2\text{O (liq.)} \rightleftharpoons \text{H}_2\text{S (gas)} + \text{HDO (liq.)}$$

the equilibrium constant is given by

$$K = \frac{[\text{H}_2\text{S (gas)}][\text{HDO (liq.)}]}{[\text{HDS (gas)}][\text{H}_2\text{O (liq.)}]} = 2.20 \text{ at } 30°\text{C}$$

Thus, deuterium concentrates in the aqueous phase and by using a multi-stage exchange column similar to a fractional distillation column it is possible to prepare almost pure deuterium oxide. This method is in fact used for the industrial production of deuterium oxide. A similar exchange reaction between atmospheric CO_2 and aqueous bicarbonate ions,

$$^{13}\text{CO}_2\text{(gas)} + \text{H}^{12}\text{CO}_3^-\text{(soln)} \rightleftharpoons {}^{12}\text{CO}_2\text{(gas)} + \text{H}^{13}\text{CO}_3^-\text{(soln)}$$

where $K = 1.014$, accounts for the fact that carbonate deposits in the Earth such as limestone are slightly enriched in ^{13}C relative to atmospheric CO_2. The distribution of isotopes in nature presents a fascinating topic for study.

Kinetic isotope effects are probably the most important in biological systems since they are a measure of the effect of isotopic substitution upon the rates of reactions. All the atoms in a molecule are in a constant state of motion and all chemical bonds possess vibrational energy. The zero-point energy of a bond is given by

$$E_0 = \tfrac{1}{2}h\nu$$

where h is Planck's constant and ν is the frequency of vibration which depends upon the masses of the atoms forming the bond, just as the frequency of vibration of a spring depends upon the weight attached to it. For example, a deuterium atom is twice as heavy as a hydrogen atom and therefore a C—D bond has a lower frequency of vibration

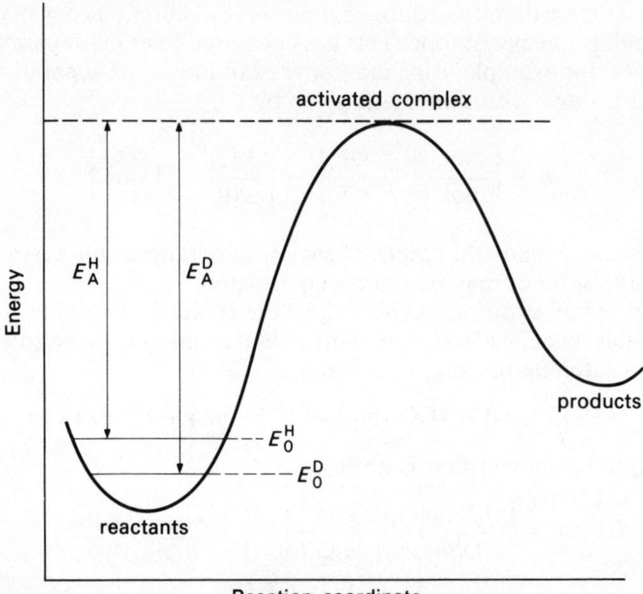

Figure 5.3 Diagrammatic representation of the origin of an intermolecular isotope effect. E_0^H and E_0^D represent the zero-point energies for C–H and C–D bonds respectively; E_A^H and E_A^D represent activation energies for scission of these bonds during the course of a chemical reaction. The higher activation energy required for scission of the C–D bond will usually cause a reduction in the rate of reaction of the deuterated compound.

and hence a lower zero-point energy than a C—H bond. In a chemical reaction, when proceeding from reactants to products, an energy barrier (the activation energy) has to be surmounted as shown in Figure 5.3. The rupture of a C—D bond requires a higher activation energy, E_A, than rupture of a C—H bond. Thus, the kinetic isotope effect manifests itself in a smaller rate constant, k_D, for C—D bond rupture and is usually expressed as the ratio of the rate constants k_H/k_D. The effect is of great value in the study of reaction mechanisms since it permits detailed examination of rate-determining steps and transition states of chemical reactions. However, particularly in the case of the hydrogen isotopes, the effect can be a very serious limitation on their usefulness as tracers since rate differences up to tenfold have been observed for deuterium and of course much higher differences are possible for tritium. For isotopes of higher mass the effect is much less serious since the mass difference is greatly reduced. For example, ^{14}C is only 17% heavier than ^{12}C and the maximum possible kinetic isotope effect, k_{12}/k_{14} for $^{12}C-^{14}C$ bond scission is around 1.15, but is usually much less than this (<1.1), and therefore not a serious problem in tracer work.

Tritium can be used with complete confidence for studying uptake of labelled compounds and even for metabolic studies provided that it is known that *no scission of the isotopically substituted chemical bond occurs*. Uncritical use of deuterium or tritium, however, can lead to totally misleading experimental results. The following examples illustrate the point:

(a) Rats given 50% D_2O in H_2O as their sole source of water die before their body fluids reach equilibrium with their water supply.

(b) When peppermint plants (*Mentha piperita* L.) are watered with D_2O/H_2O, a progressive inhibitory effect on growth is observed above 20% D_2O and at the 70% level growth is essentially stopped. The major effect appears to be inhibition of cell division.

(c) In a study of the biosynthesis of choline and creatinine from methanol in the rat, doubly labelled methanol ($^{14}CH_3OH$ + $^{12}CH_2TOH$) was used as the precursor. As the experiment proceeded it was observed that the T/^{14}C ratio in the recovered methanol had increased showing that the C—T bonds were reacting less frequently than the C—H bonds. Thus, use of the tritium-labelled precursor alone would have given an erroneous measure of the rate of biosynthesis.

An example of a kinetic isotope effect in a biological system is given in Experiment 5.6.2.

5.6 Practical experiments

5.6.1 Counting statistics

(a) Apparatus and materials *Either* (a) a thin mica end window-type Geiger–Müller counter mounted in a lead castle with source mounting shelves beneath and a standard source of carbon-14 (disc of labelled poly(methylmethacrylate)) *or* (b) a liquid scintillation counter and a standard sealed carbon-14 source.

(b) Procedure

(1) Using the GM counter set the counter high voltage to 50 V above threshold (see practical in Ch. 2). With the source in position take a series of measurements (at least 40) each of 100 s.

(2) Using a scintillation counter, set the machine to count the sample repeatedly for 100 s counts for a total of at least 40 measurements. In both cases tabulate your results in the following way:

Counts per 100 s	*Deviation from mean count* Δ_i^*	*Square of deviation,* Δ_i^2
.	.	.
.	.	.
.	.	.

$^*\Delta_i = C_i - \bar{c}$ (see p. 81).

(c) *Analysis of results*

(1) From the results calculate the standard deviation

$$\sigma = \frac{\Sigma \Delta_i^2}{(N-1)}$$

where N is the total number of observations.
(2) Check that this approximately equals the square root of the mean count.
(3) Calculate the percentage of the number of deviations exceeding the standard deviation. This should approximate to the theoretical value of 32.7%.
(4) For a Poisson distribution, mean deviation ÷ standard deviation = 0.798.
(5) Calculate the 95% confidence level limits and check that only about 5% of your results fall outside these limits.
(6) Your results will not agree exactly with the theoretical values but should not be far out. A useful statistical test, which may be used to check that the observed distribution of results approximates to a Poisson distribution, is to calculate chi-squared (χ^2):

$$\chi^2 = \frac{\sum_{i=1}^{i=N} \Delta_i^2}{N}$$

The value of χ^2 varies with the number of observations and the level of confidence required. For a confidence of 90% the limits of χ^2 are

No. of observations	*Limits of* χ^2
30	19.8–39.1
40	28.2–50.7
50	36.8–62.0

If your result exceeds the upper limit, excess variation is indicated. If the value is less than the lower limit, a 'selection' process or non-random error is occurring.

5.6.2 Experimental demonstration of a kinetic isotope effect using tritiated water

(a) Apparatus and materials

(1) 500 cm^3 conical flask
(2) Flat bed rotary shaker
(3) Czapek Dox minimal medium (contains the following per litre: sodium nitrate, 2.0 g; potassium chloride, 0.5 g; magnesium glycerophosphate 0.5 g; ferrous sulphate, 0.01 g; potassium sulphate, 0.35 g; sucrose 30.0 g; pH 6.8)
(4) Buchner flask and funnel
(5) Oven
(6) Combustion apparatus – see procedure
(7) Liquid scintillation counter
(8) Tritiated water
(9) 1 ml syringe with a sterile millipore filter attached
(10) Culture of *Verticillium albo-atrum* (or equivalent fungus).

(b) Procedure Put 200 cm^3 of Czapek Dox medium into the conical flask, seal with a firm cotton wool bung, loosely cover the top with aluminium foil and autoclave for 15 min at 15 p.s.i. (121 °C) to sterilize the medium. When cool introduce 1 cm^3 of tritiated water (1.85×10^5 Bq; 5 μCi) into the flask using sterile techniques. (First draw up the tritiated water into the syringe, attach the millipore filter apparatus to the end of the syringe and then carefully force the water through the filter into the flask. N.B. After removing and again before re-inserting the cotton wool bung into the flask, flame the neck of the flask: in addition do not allow the cotton wool bung to contact any non-sterile surface.) Gently swirl the flask to distribute and mix the tritiated water with the medium and then inoculate the mixture (again using sterile technique) with the fungus (if *Verticillium* cultures are used, then 0.5 cm square plugs from 10-day-old potato

dextrose agar cultures are suitable). Shake the culture continuously at room temperature using a rotary shaker at 150 rev min^{-1} and for 14 days.

After the growing period harvest the fungal mycelium using a buchner funnel and flask. Keep the filtrate for measurement of the specific activity of tritiated water in the growing medium. Carefully transfer the filtered mycelium to a small glass dish or watch glass and place in an oven (in a fume cupboard) and dry the material at 105°C overnight. The mycelium must now be burnt and the water from the combustion collected and its specific activity measured. Any available means of dry combustion may be used. One convenient method is to use a quartz tube approximately 40 cm long by 1.5 cm diameter which contains a short length of copper oxide wire ($\simeq$7 cm) packed at one end. This end of the tube is then attached to a small pre-weighed trap or 'U' tube immersed in an ice-salt freezing mixture. The open end of the trap is protected by an anhydrous magnesium perchlorate drying tube. A sample of mycelium ($\simeq$0.15 g) is weighed into a small boat and placed roughly in the centre of the quartz tube. A slow stream of dry air is passed through the tube. The copper oxide is heated using the hottest flame of a bunsen burner (nearly red heat). A second bunsen is used to heat the mycelium, gently at first and then more strongly as thermal decomposition occurs. Finally, all the charred material and brown tars are 'chased' down the tube into the copper oxide. Water from the combustion is collected in the trap, weighed and then transferred to a counting vial by repeated washings using small quantities of a suitable liquid scintillant (total 10 cm^3).

Perform several combustions and then compare the mean specific activity of the water with the mean specific activity of the filtered growth medium determined using the same scintillator. The isotope effect may be expressed in terms of the relative specific activity:

$$\frac{\text{specific activity of combustion water}}{\text{specific activity of culture medium}}$$

N.B. The magnitude of the isotope effect as measured in this experiment is slightly larger than might be expected as a consequence of some dilution of the tritium by hydrogen atoms from the sucrose used in the culture medium. However, most of these hydrogen atoms will have become equilibrated with the tritiated water during the experiment and only a small proportion of them will be incorporated into the fungus without exchange. An estimation of this non-exchanged hydrogen would be fairly difficult to obtain and would require several additional experiments to ascertain.

6 Special analytical methods used in biological research

6.1 Introduction

When radioisotopes are used as tracers in biological systems they are inevitably being used in some form of analysis. Their unique properties permit extremely sensitive determination of the fate of individual atoms in molecules, often in a non-destructive manner. In this chapter several special techniques of analysis are described and examples of their application given.

6.2 Isotope dilution analysis and isotope derivative analysis

6.2.1 Isotope dilution analysis

When it is necessary to measure quantitatively one compound in a mixture of very similar compounds (e.g. sugars, amino acids or fatty acids) conventional methods often fail because quantitative isolation and high purity are mutually exclusive concepts. Isotope dilution, however, enables the analyst to pursue high purity at the expense of total recovery and yet to obtain a quantitative measure of the compound in question. The basic method is very simple and depends upon the fact that if a labelled compound is mixed with the corresponding unlabelled compound the specific activity will be reduced by an amount determined by the dilution factor. Thus, to analyse for an unknown quantity m_i moles of an inactive compound A in a mixture of compounds A + B + C + D + . . . a pure radioactive form of the same compound A^* is required. The analytical procedure is then as follows:

(1) Measure the specific activity of A^*, i.e. S_0 Bq mol^{-1}.
(2) Add m moles of A^* to the mixture to be analysed and ensure that total mixing is achieved, preferably in homogeneous solution.
(3) Isolate a sample of $A + A^*$ from the mixture by whatever

means possible and purify repeatedly until the specific activity of the sample, S Bq mol^{-1}, is constant.

Note that the isolation need not be quantitative and that purification can be as wasteful as is necessary to obtain high purity. All that is required is sufficient sample for radioactive assay.

The calculation of the unknown m_i is then as follows. Total activity added, a Bq, is given by

$$a = mS_0$$

the total activity remains unchanged but, after dilution,

$$a = (m + m_i)S$$

therefore

$$mS_0 = (m + m_i)S \tag{6.1}$$

or

$$m_i = m[(S_0/S) - 1]$$

This simple technique is most accurate when there is an appreciable difference in the two specific activities, i.e. when $m_i \gg m$, provided that the count rate of the diluted sample is large enough to be determined accurately. It is also an ideal method for determination of total quantities in biological pools. For example, in live animal studies total sodium ions can be determined by injecting a small amount of ^{24}Na and then sampling blood serum from time to time. Similarly, tritiated water can be used for total body water determination.

6.2.2 Reverse isotope dilution analysis

Frequently, the use of tracers in biological experiments results in complex mixtures of labelled compounds. A given radioactive product A* in such a mixture can be determined quantitatively using reverse isotope dilution. Two procedures are possible depending upon whether or not the specific activity S_0 of A* is known. (S_0 can only be known if A* was the starting material which has been partially metabolized or if the chemical or metabolic pathway between A* and the starting material is clearly defined.)

(a) The specific activity of A* is known In this case a small portion of the radioactive mixture is added to a relatively large known amount m_i of pure inactive A (carrier) and the mixture completely homogenized. Then A + A* is rigorously purified to remove all the other radioactive constituents. As before, this process can be as wasteful as necessary to obtain constant specific activity S. The calcu-

lation is exactly as before except that in this case m is the unknown quantity and therefore equation 6.1 is rearranged to give

$$m = \frac{m_i}{(S_0/S) - 1} \tag{6.2}$$

This method is one of the preferred ways of determining the radiochemical purity of labelled compounds (see p. 90) and is frequently quoted in manufacturers' specifications relating to their radiochemicals. The procedure is slightly modified in that the radiochemical is accurately diluted with carrier and the specific activity S_u of the diluted material determined before purification. Then after extensive purification (which may involve recrystallization from several different solvents and/or several chromatographic purifications) the specific activity S is determined. The radiochemical purity is then expressed as

$$\frac{S}{S_u} \times 100\%$$

(b) The specific activity of A^* is not known In this case there are two unknowns, m and S_0, and the technique is to take two portions of the radioactive mixture and dilute to different extents with quantities m_i and m_i' respectively of the carrier compound A. The method is often referred to as isotope double dilution analysis and results in two diluted samples which after purification have specific activities S and S'. It is then possible to solve a pair of simultaneous equations,

$$mS_0 = (m + m_i)S$$
$$mS_0 = (m + m_i')S'$$

to give

$$m = \frac{m_i'S' - m_iS}{S - S'} \tag{6.3}$$

Clearly, with double the experimental operations and radioactive assays, the precision of this method is not as high, but it is extremely useful in situations where the product is the result of complex biochemical pathways such as photosynthesis from $^{14}CO_2$ or metabolism of [^{14}C]acetate.

6.2.3 Isotope derivative dilution analysis

For mixtures of inactive compounds where the quantity m_i of A is small, simple isotope dilution becomes inaccurate due to the similarity of S_0 and S. (The factor $(S_0/S) - 1$ becomes subject to large errors when we are attempting to measure a small difference between two large numbers.) The dilemma is resolved by using a high specific activity radioactive reagent R* which will react quantitatively with A to form a radioactive derivative AR*. The quantity of AR* present in the mixture can then be determined by reverse isotope dilution as described in Section 6.2.2(a), using unlabelled derivative AR as carrier. The classic example of the use of this method was the determination of individual amino acids in protein hydrolysates by reaction with *p*-[^{131}I]iodobenzenesulphonyl chloride (pipsyl chloride):

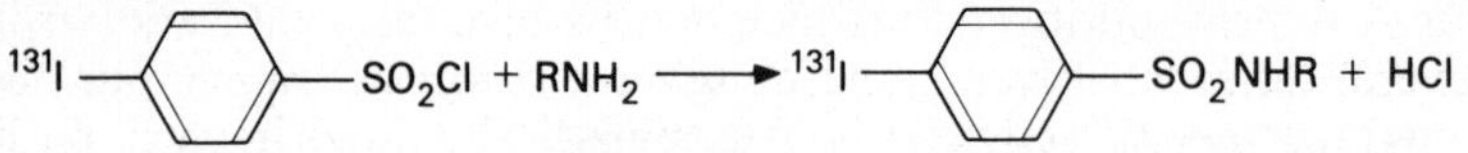

$$^{131}I-C_6H_4-SO_2Cl + RNH_2 \longrightarrow {}^{131}I-C_6H_4-SO_2NHR + HCl$$

The precision of the method is limited only by the counting errors (±1–2%).

6.2.4 Double isotope derivative analysis

So far, in the methods described, the non-quantitative aspect of purification has been stressed. Nevertheless, there is a limit to this since to determine the specific activity S it is necessary to weigh or quantify in some way the sample of A + A* whose radioactivity is measured (since specific activity is defined as activity per unit weight, see p. 21). A further refinement which overcomes this limitation and greatly increases the sensitivity is as follows. After preparation of the [^{131}I]pipsyl derivatives of the amino acids in the unknown mixture, a known quantity of ^{35}S-labelled pipsyl derivative

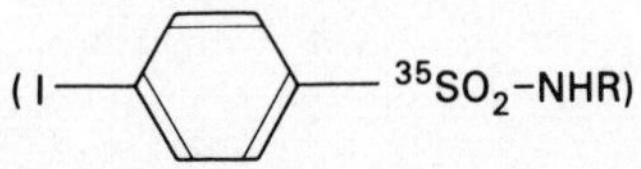

$$(I-C_6H_4-{}^{35}SO_2-NHR)$$

of the particular amino acid under assay was added. This acted as a recovery indicator and thus the weight of amino acid in the original sample could be accurately determined from ^{35}S/^{131}I counts ratio in a radiochemically pure specimen of the derivative. An even greater improvement was achieved by adding a ^{14}C-labelled sample of the particular amino acid to the mixture before derivative formation. This had the added advantage that it was no longer essential for quantitative formation of derivative as the ^{14}C content was a measure of all losses both during the reaction and the subsequent isolation. Using similar notation as before: m_i is the number of moles of A in the unknown mixture, m is the number of moles of radioactive A* containing isotope 1 added to the mixture, S_0 is the specific activity of A*,

a is the total activity of A^* ($= mS_0$), S_r is the specific activity of reagent R^{**} containing isotope 2. After formation of the derivative $(A + A^*)R^{**}$, purification may be effected in any suitable way, unweighed amounts of inactive carrier AR being added where and as many times as necessary until a radiochemically pure sample is finally obtained. Then, if a_1 is the activity due to isotope 1 in the assay sample, the factor to correct for all losses during preparation and purification of the derivative will be

$$\frac{a}{a_1} = \frac{mS_0}{a_1}$$

If a_2 is the activity due to isotope 2 in the assay sample, the total activity a_r due to the second isotope in the derivative is given by

$$a_r = a_2 \frac{mS_0}{a_1}$$

also

$$a_r = m_i S_r$$

(since m_i moles of A give m_i moles of AR^*); therefore

$$m_i = m \frac{a_2 S_0}{a_1 S_r} \tag{6.4}$$

This simple equation represents a very powerful analytical tool for the measurement of minute traces of amino acids, proteins, peptides, hormones and drugs in biological systems. With modern liquid scintillation counters capable of accurate assay of doubly labelled samples containing ^3H and other isotopes, with high resolution chromatographic techniques and very high specific activity tritium-labelled reference compounds, it is now possible to measure quantities as small as 10^{-10}–10^{-11} g. Many examples are described in the literature cited in Appendix D.

6.2.5 Substoichiometric isotope dilution analysis

Another ingenious method of dealing with the problems associated with very small quantities has been described by Ruzicka and Stary (1961). Their method also avoids the necessity of determining the mass of the diluted samples used for activity measurement. A known amount of labelled A^* is mixed with A as before and the mixture allowed to form a derivative with a reagent which is present at less than the necessary equimolar amount (i.e. a substoichiometric amount of R is used). Thus, A and A^* react with the reagent in the exact proportions determined by the dilution factor. An exactly equal substoichiometric amount of R is reacted with pure A^*. The two samples are then isolated and counted and the ratio of their activities is equal to the ratio S_0/S required for substitution into equation 5.1. The 'reagent' used in this method really need only be some method for isolating exactly equimolar amounts of $(A + A^*)$ and A^*. Thus, as well as true derivative-forming reagents, electrolysis, solvent extraction, precipitation and adsorption methods have all been reported. The method has been used mostly for inorganic analyses, particularly in association with radioactivation analysis (Sec. 6.4) and amounts as low as 10^{-10}–10^{-11} g detected. A similar procedure of very wide application in biochemical analysis is described in the next section.

6.3 Radioimmunoassay

Radioimmunoassay is the name given to an ever increasing and extremely important collection of methods for the assay of a wide variety of compounds ranging from peptide hormones of all kinds (e.g. insulin, growth hormone, gastrin) to non-peptide hormones (e.g. cortisone, testosterone, thyroxine) to non-hormonal substances such as digoxin, albumin, morphine, LSD and many others. The procedures have become commonplace and routine in most clinical and veterinary laboratories throughout the world and there is a huge market in 'kits' for the determination of specific substances. Some of the methods have been termed 'saturation analysis' since saturation of active sites on an antibody is involved. The term 'binding assays' has also been suggested, particularly as some important assays make use of non-immune specific binding reagents; however, the procedure which most nearly describes the general method is *substoichiometric isotope dilution*, as described in the last section. The methods are based on the pioneering work of Berson and Yalow who published their first results in 1959. Briefly, to analyse for an unknown quantity of a compound in a solution, a known quantity of the radioactive compound is added and the mixture incubated with a substance which will specifically bind to that compound. The concentration of the 'binding molecule' is adjusted so that only a limited number of binding sites are available and the labelled and unlabelled compounds must compete for binding. Subsequently, the 'bound' and 'free' fractions are separated and either one or both are radioassayed.

Separation is achieved in several ways, depending on the substances, but methods include electrophoresis, dialysis, gel filtration, adsorption and precipitation. The most convenient methods are those which can be processed in large numbers with a minimum of manipulation. Thus, precipitation with centrifugation or adsorption on charcoal are popular. When a new analytical method is being set up, however, it is wise to use more than one type of separation initially in order to check their validity.

Because biologically produced antibodies or specific binding reagents cannot be defined in the same way that conventional analytical reagents can, it is not possible to carry out the exact substoichiometric procedure. Instead a calibration curve is prepared using known amounts of 'antigen' and a standardized incubation and separation procedure as set out in Figure 6.1.

For assay of any specific substance two requirements have to be met. Firstly, a radioactive substrate must be available which is identical with the substance to be measured, or at any rate so similar that the 'binding molecule' will recognize it. Secondly, the 'binding molecule' must be of adequate specificity and high affinity if the assay is to be sufficiently sensitive.

In addition to the above, for adequate sensitivity the radioactive substrate must have a very high specific activity. In practice this means either tritium labelling for steroids, drugs, etc., or iodine-125 labelling for proteins (occasionally ^{75}Se and ^{57}Co are used in special cases). Iodine reacts with the tyrosyl residues in proteins and the labelling procedure is fairly simple (although potentially hazardous unless great care is exercised); however, provided that this procedure does not denature the protein, the iodinated product is usually acceptable for reaction with a specific antibody.

Antisera are raised by exposure of experimental animals to the antigen. Proteins are usually naturally immunogenic but smaller molecules like digoxin and steroids have to be conjugated with a protein such as bovine serum albumin before suitable antibodies can be raised. The methods used tend to be empirical and are outside the scope of this book.

The development of radioimmunoassay represents a major contribution to the field known as nuclear medicine. Using standard kits hospital laboratories can now carry out routine analyses at the picogram level for a wide variety of materials affecting health, medical disorders and disease. Such analyses were impossible before the advent of radioisotope techniques and their development is an excellent example of the benefits to man emanating from the study of nuclear science.

6.4 Radioactivation analysis

When stable nuclides are bombarded with nuclear projectiles it is possible to induce nuclear reactions which are used to produce

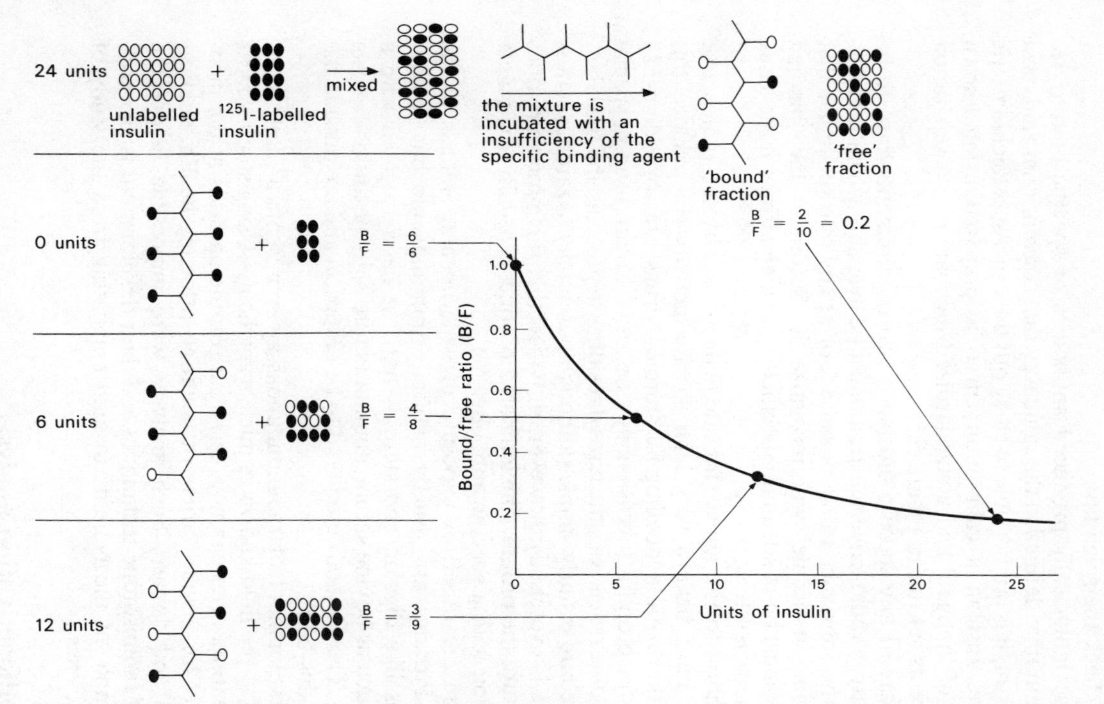

Figure 6.1 Diagrammatic representation of the radioimmunoassay of insulin, showing how a calibration curve, relating the 'bound free' ratio to the concentration of unlabelled insulin in serum, is prepared from standard solutions containing known amounts of insulin. Each solution is mixed with a fixed amount of labelled insulin and then the mixtures are treated with sub-stoichiometric amounts of specific binding agent. The bound and free fractions are separated and the radioactivities determined.

radioactive isotopes (see Ch. 1, p. 2). The same nuclear reactions also form the basis of an important and extremely sensitive method of trace element analysis known as radioactivation analysis. The simplest and most widely used technique is to induce radioactivity in the sample to be analysed by irradiation with neutrons in a nuclear reactor. Charged nuclear particles such as protons, deuterons and α-particles may also be used as projectiles and high energy γ-rays can induce some nuclear reactions, but their use is rather specialized. Provided that suitable conditions are selected, almost every element in the periodic table is susceptible to analysis by using neutron bombardment. The most common nuclear reaction is the (n, γ) reaction and a few selected examples are listed in Table 6.1. The method is clean, simple, very specific, remarkably free from major sources of error and is often non-destructive. It is also possible to achieve high sensitivities (ranging from 10^{-6} to 10^{-12} g) and thus, for example, to analyse for trace elements in samples such as a single nerve fibre or a fragment of a single hair.

The standard procedure is to select a sample for analysis and prepare a standard which is as similar to the sample as possible and which contains a known amount of the element to be measured. The sample and standard are then irradiated together in a nuclear reactor (most universities now have access to this facility). The length of time of irradiation depends on various factors (discussed in more advanced texts) but in general there is little to be gained by irradiating for more than four times the half-life of the generated radioactive nuclide of interest.

After irradiation the sample and standard are removed and their activities determined. In most cases the nuclide formed will be a γ-emitter and analysis may be carried out directly using a high resolution detector (usually of the semiconductor type) and a multi-channel

Table 6.1 Some examples of thermal neutron (n, γ) reactions used in radioactivation analysis.

Element	*Reaction*	*Half-life of product nuclide*	*Sensitivity (g)*
iron	$^{58}Fe(n, \gamma)^{59}Fe$	45 d	1×10^{-7}
copper	$^{63}Cu(n, \gamma)^{64}Cu$	12.8 h	1×10^{-10}
zinc	$^{68}Zn(n, \gamma)^{69m}Zn$	13.8 h	5×10^{-9}
arsenic	$^{75}As(n, \gamma)^{76}As$	26.5 h	5×10^{-11}
gold	$^{198}Au(n, \gamma)^{199}Au$	2.7 d	5×10^{-12}

pulse height analyser. Provided that there are no interfering radiations of the same energy the ratio of the activities of sample and standard are determined by measuring the ratio of the counts in a very narrow channel covering the appropriate γ-ray photopeak. The unknown weight x of the element E may then be calculated from the following equation:

$$x = \text{weight of E in standard} \times \frac{\text{net count rate of sample}}{\text{net count rate of standard}} \quad (6.5)$$

If the formed nuclide is a β-emitter with no γ-rays (e.g. ^{45}Ca), then it is necessary to separate the element chemically for assay. This is usually done by adding carrier and using standard chemical procedures to isolate the material in a pure form for counting.

By a happy coincidence, the most common biological elements are the light elements, C, H, N, O, P and S, which are not activated by low energy (thermal) neutrons. Thus, trace element analysis in biological samples can be carried out without interference from these more abundant elements.

Radioactivation has been used extensively in environmental analysis for identifying sediments in water, aerosols in the atmosphere, sources of pollution, etc. Uses in forensic science include matching hair samples, matching bullet traces on bone, determining antimony and barium on the hands of people who have fired guns, and tracing the origins of narcotic drugs. Other biological applications include measurement of trace elements in blood cells, mercury in fish, copper and manganese in brain tissue, pesticides in grain and other crops, and copper and arsenic in wine.

Finally, there are two very general applications. Firstly, radioactivation analysis is often so specific and free from complications due to impurities that it may be used to check and to calibrate other less sensitive means of analysis such as spectrophotometry or colourimetry which may then be used for routine determinations. Secondly, it may be used as a means of assay for non-radioactive tracer techniques. Thus, if use of radioactive materials is prohibited for health hazard, environmental pollution or other reasons, a stable isotope may be incorporated into a system and its presence finally detected at the end of the experiment by radioactivation.

6.5 Double labelling

In some experiments a great deal of additional useful information can be obtained if two isotopes are used instead of one. For example, it may be desired to measure the synthesis of both protein and nucleic acid in a single experiment. This could be done by using, say, a ^{14}C

precursor for the protein and a ^{3}H precursor for the nucleic acid. Tritium isotope effects can sometimes be detected using $^{3}H/^{14}C$ double labelled compounds (see p. 96) and isotope dilution analyses are often facilitated by use of a second isotope as a 'recovery indicator' (see p. 104). However, use of two isotopes does complicate the assay procedure and can also lead to increased counting errors, particularly with β-particle emitters such as ^{3}H, ^{14}C and ^{32}P since there is always overlapping of the particle energy spectra. With nuclides emitting γ-rays the situation is much simpler, provided that the photon energies are reasonably different, since each may be counted in a narrow window of a pulse height analyser (see p. 57) without interference from the other. With nuclide mixtures incorporating pure β-emitters it is necessary to use a counting method which will either account for or remove any possible overlap. Various methods are available depending on the isotope mixtures being counted, and on the equipment available. These are briefly considered in the following pages.

6.5.1 Counting separate regions of the pulse height spectrum in a liquid scintillation counter

This method depends upon there being an adequate separation of at least parts of the pulse height spectra of the two isotopes in the mixture. Common isotope pairs for which this technique is used are ^{3}H and ^{14}C, ^{3}H and ^{32}P, and ^{14}C and ^{32}P. Figure 6.2 (a and b) shows typical spectra for ^{3}H and ^{14}C indicating the extent of overlap observed with quenched and unquenched samples. It is clear from this figure that for mixtures of ^{3}H and ^{14}C it is of paramount importance that the extent of quenching is accurately determined. The method of choice is usually that of automatic external standard channels ratio (AESCR). In this method two channels are used for counting sample activity, plus two channels for external standardization. Suppose that we wish to measure the activities of a mixture of ^{14}C and ^{3}H in a series of samples, that the disintegration rates of the two isotopes are expected to vary independently, and that some of the samples might contain zero activity of the isotopes. The following technique could be used.

One channel (channel 1) in the counter is optimized for the counting of ^{14}C with the lower discriminator set to exclude practically all of the counts (>95%) arising from ^{3}H. A second channel (channel 2) is optimized for counting ^{3}H. This channel will also contain counts arising from ^{14}C because of the spectral overlap: the higher the quenching the higher the overlap. Two other channels (channels 3 and 4) are used to detect the lower and upper halves of the external standard pulse height spectrum respectively for assessment of counting efficiency using the AESCR method (see p. 60).

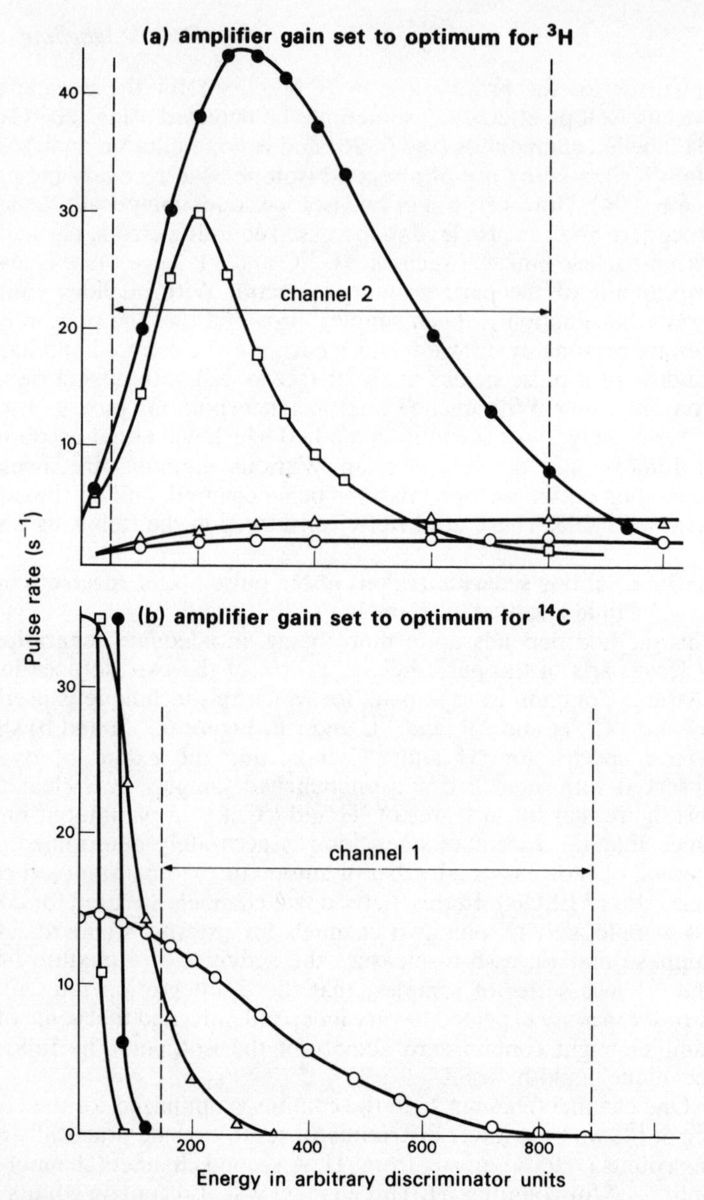

Figure 6.2 (a) Typical spectra for unquenched and quenched samples of ^{3}H and ^{14}C. Channel 2 of the counter has been optimized for measuring ^{3}H: this channel will also contain counts arising from ^{14}C because of the spectral overlap. (b) Typical spectra for unquenched and quenched samples of ^{3}H and ^{14}C. Here channel 1 of the counter has been optimized for measuring ^{14}C with the lower discriminator set to exclude most of those counts arising from the unquenched ^{3}H sample. ○, ^{14}C unquenched; △, ^{14}C quenched; ●, ^{3}H unquenched; □, ^{3}H quenched.

In order to standardize the counter, two sets of single-labelled quenched standards are required, one containing ^{14}C and the other ^{3}H. These standards are used to construct the quench correction curves: (a) ^{14}C efficiencies in channel 1 = E_1; (b) ^{14}C efficiencies in channel 2 = E_2; (c) ^{3}H efficiencies in channel 2 = E_3 (Fig. 6.3). The three efficiency curves are then used to calculate the absolute ^{14}C and ^{3}H disintegration rates as follows:

$$^{14}\text{C disintegration rate} = \frac{^{14}\text{C counts per minute in channel 1}}{E_1} \tag{6.6}$$

$$^{14}\text{C counts per minute in channel 2} = {}^{14}\text{C disintegration rate} \times E_2 \tag{6.7}$$

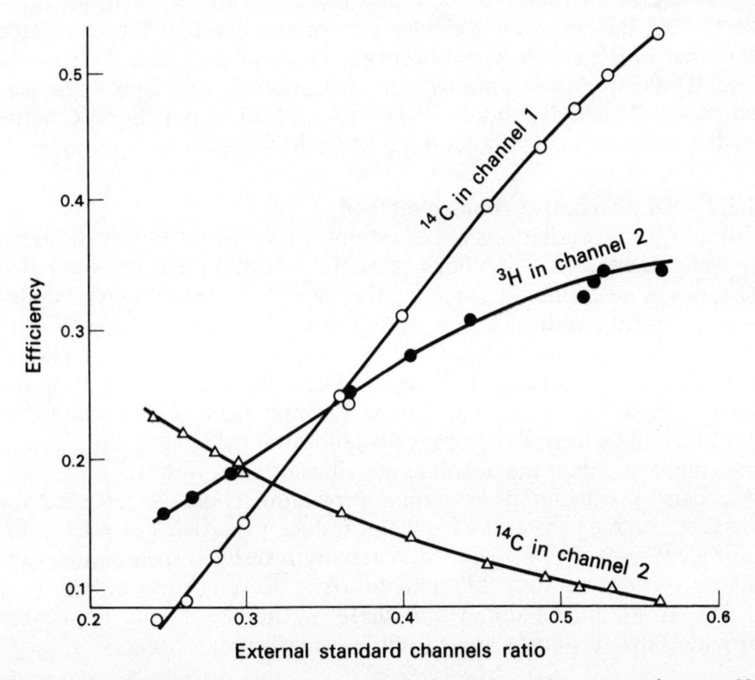

Figure 6.3 Efficiency/external standard channels ratio curves (see p. 60) for ^{14}C and ^{3}H samples. Note that as the efficiency of counting decreases (i.e. quenching increases) an increasing proportion of the ^{14}C is measured in the ^{3}H channel (see also Fig. 6.2a).

$$^{3}\text{H counts per minute in channel 2} = \text{total counts per minute in channel 2} - {}^{14}\text{C counts per minute in channel 2} \quad (6.8)$$

$$^{3}\text{H disintegration rate} = \frac{^{3}\text{H counts per minute in channel 2}}{E_3} \quad (6.9)$$

In principle, this method can be used for measuring the activities of mixtures of any two α-emitting or β-emitting isotopes providing that the spectral overlap is not too great. Care has to be taken in some mixtures, particularly at low counting efficiencies, because the increased quenching of the higher energy isotope into the lower channel increases the uncertainty of calculations concerning the relative contributions each isotope makes to the counts in this channel.

A full consideration of the errors inherent in double label counting is beyond the scope of this book, but, as far as ^{3}H and ^{14}C mixtures are concerned, where the counting efficiency of ^{14}C is about three times that for ^{3}H, then a similar percentage error in the calculated activities of the two isotopes occurs at ratios of the ^{14}C disintegration rate/^{3}H disintegration rate between 0.1 and 0.5; that is, for practical purposes, ^{3}H activity should be two-to-ten fold higher than ^{14}C activity for optimal precision counting of both isotopes.

6.5.2 An alternative counter method

Where emitted radiations are of different natures or of very different energies then it is sometimes possible to distinguish between two isotopes by counting the sample either in different machines or under very different conditions. For example, a mixture containing ^{14}C and ^{125}I isotopes could be assayed first by counting total activity in a liquid scintillation counter (gives ^{14}C and ^{125}I counts) and then counting the same sample in a γ-counter (gives ^{125}I counts only). Providing that both machines have been properly calibrated using separate ^{14}C and ^{125}I standards, then the absolute activities of each isotope in the mixture can be calculated. A similar procedure could be adopted for mixtures such as ^{14}C and ^{32}P which produce β-particles of widely differing energies. In this case, total activity is derived from liquid scintillation counting and ^{32}P activity from Cerenkov counting (see p. 61). It should be noted that these methods are only applicable provided that replicate assay samples are all very similar.

6.5.3 Decay method

Where two isotopes that have widely differing decay rates are being counted then it is possible to use this approach provided that one of the isotopes has a half-life which is very long in comparison to the period over which measurements are made. For example, suppose that we have a mixture of ^{3}H and ^{32}P, then the following considerations would apply.

At the time ($t = 0$) of initial measurement

$$\text{cpm}_{(0)} = \text{cpm}_{^{3}\text{H}} + \text{cpm}_{^{32}\text{P}_0} \tag{6.10}$$

(cpm refers to the number of counts per minute). At a time some days later ($t = t$) when the ^{32}P has measurably decayed but the ^{3}H negligibly so, then

$$\text{cpm}_{(t)} = \text{cpm}_{^{3}\text{H}} + \text{cpm}_{^{32}\text{P}_t} \tag{6.11}$$

Hence

$$\text{cpm}_{(0)} - \text{cpm}_{(t)} = \text{cpm}_{^{32}\text{P}_0} - \text{cpm}_{^{32}\text{P}_t} \tag{6.12}$$

From a consideration of the decay equations we have

$$\text{cpm}_{^{32}\text{P}_t} = \text{cpm}_{^{32}\text{P}_0}\, e^{-\lambda t} \tag{6.13}$$

Substituting for $\text{cpm}_{^{32}\text{P}_t}$ in equation 6.13 and rearranging, then

$$\text{cpm}_{^{32}\text{P}_0} = \frac{\text{cpm}_{(0)} - \text{cpm}_{(t)}}{1 - e^{-\lambda t}} \tag{6.14}$$

Having found $\text{cpm}_{^{32}\text{P}_0}$, then $\text{cpm}_{^{3}\text{H}}$ is obtained by difference.

6.5.4 Separation methods

Provided that the two isotopes of interest occur in separate compounds and that no cross-contamination of label occurs, then any method which allows the compounds to be separated (e.g. chromatography) can be used to separate the nuclides which are then assayed independently. Should it not be possible to separate the components of a mixture or if the two isotopes are contained within the same compound then some chemical means of separation must be used before any assessment of radioactivity is made. For mixtures containing ^{14}C and ^{3}H the simplest method of separation is to burn the sample and collect and measure the $^{14}CO_2$ and $^{3}H_2O$ evolved separately. If only a few samples are to be assayed, combustions are conveniently carried out by the Schöniger flask method, but for large numbers commercially available sample oxidizers are generally used (see p. 66).

If it is not possible to use a sample oxidizer, other alternative chemical methods to separate the radiolabelled components of interest must be used before measurement. The means used for this obviously depend upon the compound or radioisotopes being measured and so no definite directions are possible.

6.6 Radiochromatography

The application of chromatography is of great significance in biological science and in combination with radioisotopes provides a wide range of important analytical techniques. However, discussion of chromatographic methods is outside the scope of this book. The assay of radioactivity on chromatograms or in chromatography eluates is covered in various other parts of the book, for example autoradiography for TLC or paper chromatograms (p. 70, see also 6.7 below). Liquid scintillation counting is widely used for counting strips cut from paper (p. 67), adsorbant scraped from TLC (p. 70), fractions taken from liquid chromatography and samples condensed from the effluent in gas chromatography. There also exist a variety of specialised instruments for scanning thin layer and paper chromatograms. These are usually based on windowless flow GM counters (p. 47) where the counting efficiencies tend to be rather low ($\simeq$10–15% for ^{14}C and <2% for ^{3}H). Nevertheless a great deal of very useful information on radiochemical purities and the identities of various metabolites can be obtained using such instruments. Continuous detection of radioactivity in liquid or gas chromatographic eluates also requires special instrumentation. Liquid effluents are usually passed through a cell filled with solid scintillator (e.g. anthracene crystals, plastic or glass scintillators) which is viewed by two photomultiplier tubes, and some manufacturers provide kits for modifying liquid scintillation counters for this purpose. Gas chromatography eluates containing ^{14}C and ^{3}H are usually passed through combustion/reduction furnaces and the resulting $^{14}CO_2$ and $^{3}H_2$ gases measured in a gas flow proportional or GM counter. Whatever method is used, comparison of the resultant twin recorder traces relating to the mass and the radioactivity of each component provides a great deal of useful information for the analyst.

6.7 Autoradiographic techniques

6.7.1 General methods

There are currently four methods commonly used for the preparation of autoradiograms from biological material.

(a) Apposition This is perhaps the simplest method. The specimen (e.g. a piece of labelled tissue or even a whole animal or plant) is

dried and held against the emulsion (usually an X-ray film) in darkness for an appropriate length of time. A good 'rule of thumb' when using X-ray film is that a specimen giving 1000 counts min^{-1} on a thin end window GM tube will require approximately 24 h exposure – variations from this will require a proportional reduction or increase in exposure time. After exposure the film is developed and the distribution of the radioisotope in the specimens is determined by exactly re-positioning the developed film on the specimen, when patterns of black opposite a tissue, organ or area specify the radioisotope location. This method is particularly useful for the location of radioactive compounds separated by chromatography and has recently been developed for the location of very low activities on chromatograms (see Section 6.7.2(a)).

(b) Mounting sections directly on an emulsion This is really a modification of the apposition method. In this case a thin section of the labelled tissue is placed directly on the emulsion for the appropriate length of time. This tissue remains on the emulsion and may be stained while still in place to aid in identifying the area containing the isotope. The disadvantage with this method is that the microscopic level of the tissue and the emulsion are different and so cannot be brought into focus at the same time. However, the method does provide for the section and the autoradiogram to be on the same slide.

(c) Stripping film method Resolution of the histological section in nearly the same plane as the autoradiogram can be achieved by this technique. In this method a section of the labelled tissue is covered with a very thin sheet of emulsion (obtained commercially from Kodak: fine grain autoradiographic stripping plate AR 10). As the emulsion dries the film contracts against the tissues of the section so that intimate contact between the source of the radioactivity and the emulsion is maintained. In this way, because the planes of the emulsion and tissue are so close, the precise location of the isotope within parts of the cells and within the large organelles of cells can be determined easily. Figure 6.4 shows the location of newly synthesized DNA in bean nuclei visualized using this technique.

(d) Liquid emulsion dipping method This technique can improve the resolution further still and is especially useful for the detection of tritium whose beta particles travel only 1 or 2 µm from their source. In this case the section mounted on a slide is coated with a thin layer of liquid emulsion by simply dipping the slide in it momentarily. The

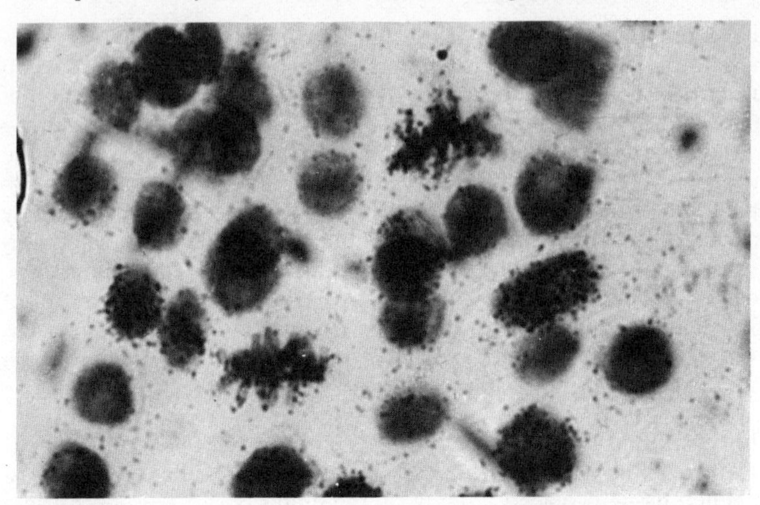

Figure 6.4 Bean roots were fed with ^{3}H-labelled thymidine and a Feulger-stained squash preparation made. Following autoradiography using the stripping film method, nuclei which have been actively synthesizing DNA are clearly located by the presence of the black dots. (Photograph by courtesy of Professor P. B. Gahan, Biology Department, Queen Elizabeth College.)

emulsion flows around the tissue before setting and, because it is therefore in close proximity, gives good resolution. As with the stripping film method, the tissue sections may be stained after exposure and development of the photographic emulsion, thus assisting in the identification of position within the tissue.

The emulsions used for microscopic autoradiography are normally especially sensitized to α- or β-particles and have small well spaced silver bromide grains for high resolution. The determination of exposure time for any emulsion is largely empirical and experimenters normally prepare several samples of their specimen and expose them for varying time intervals. In this way the optimum exposure time which gives good resolution without a high background level can be determined.

6.7.2 Special methods

(a) Fluorography In recent times the direct apposition method has been modified to meet the need to locate with increased sensitivity and speed the position of radiolabelled compounds separated by var-

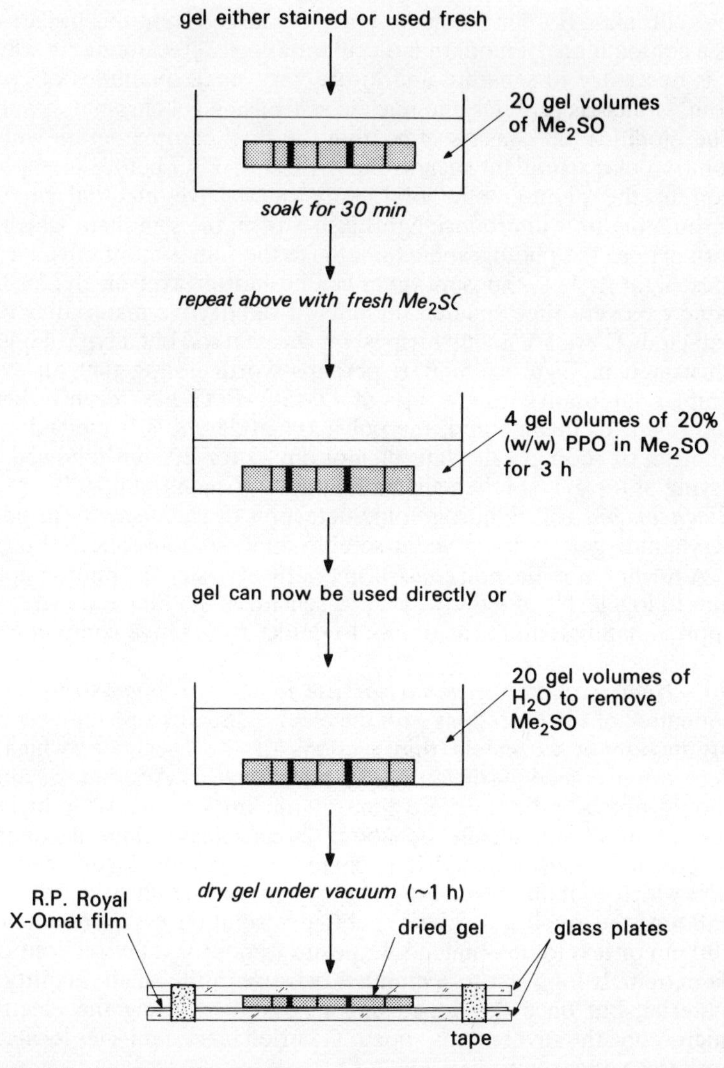

Figure 6.5 Diagram illustrating a typical procedure followed to produce a fluorogram. A disintegration rate of 3000 disintegrations min^{-1} ^{3}H per band requires about 24 h exposure; a disintegration rate of 500 disintegrations min^{-1} ^{3}H per band requires about 1 week exposure. The exposure times required can be shortened by pre-exposing film to a single flash from an electronic flash unit through suitable filters (see Laskey, R. A. and A. D. Mills 1975. Quantitative film detection of ^{3}H and ^{14}C in polyacrylamide gel by photography. *Eur. J. Biochem.* **56**, 335).

ious chromatographic methods. To a very large extent this has arisen as a consequence of modern molecular biological techniques in which it is necessary to separate and locate very small quantities of compounds such as proteins and nucleic acids using gel chromatography. The modification consists of coating the final chromatogram with a solution of a scintillant such as butyl-PBD or PPO before juxtaposition to the photographic film. Any radioactive material on the chromatogram will produce light flashes from the scintillant which in turn expose the photographic film. Since the film is more sensitive to these light flashes, exposure times can be shorter; conversely, for the same exposure time smaller amounts of radioactive material can be detected. There are many 'recipes' for this method but a typical one is illustrated in Figure 6.5. It is perhaps worth noting that an even simpler alternative to the use of DMSO–PPO has recently been described. In this method, the polyacrylamide gel is immersed in a solution of sodium salicylate (1 mol dm^{-3}) for 20 min followed by drying of the gel and then autoradiography (Chamberlin 1979: *Anal. Biochem.* **98**, 132, Fluorographic detection of radioactivity in polyacrylamide gels with the water-soluble fluor, sodium salicylate).

A further modification consists in briefly exposing the photographic film to low levels of light, which pre-sensitizes the film and increases approximately tenfold the ability to detect radioactive compounds.

(b) High resolution methods for use in electron microscopy Examination of tissue sections with the electron microscope requires the production of extremely thin sections of the specimen, which is very often embedded in an epoxy resin such as Araldite. Sections should normally be no more than 60 nm thick and for the highest resolution should ideally be about 20 nm thick. Once a suitable section has been obtained it is coated with a monolayer of emulsion which contains silver bromide crystals with grain sizes less than 150 nm, the small grain size enabling resolution of approximately 100 nm or less to be obtained. Exposure times for these sections can be extremely long (up to 3 months) because of the small quantity of material, but once developed and re-examined using the electron microscope the silver grains appear as curled black squiggles localized over the radioactivity (see Fig. 6.6).

It should be noted that because of the extreme thinness of the sections and emulsions tritium is the isotope of choice for this work since the higher energy radiations from other isotopes either pass through the plane of the ultrathin section without registering, or register their presence at a point distant from their origin, having passed obliquely through the section (Fig. 6.7).

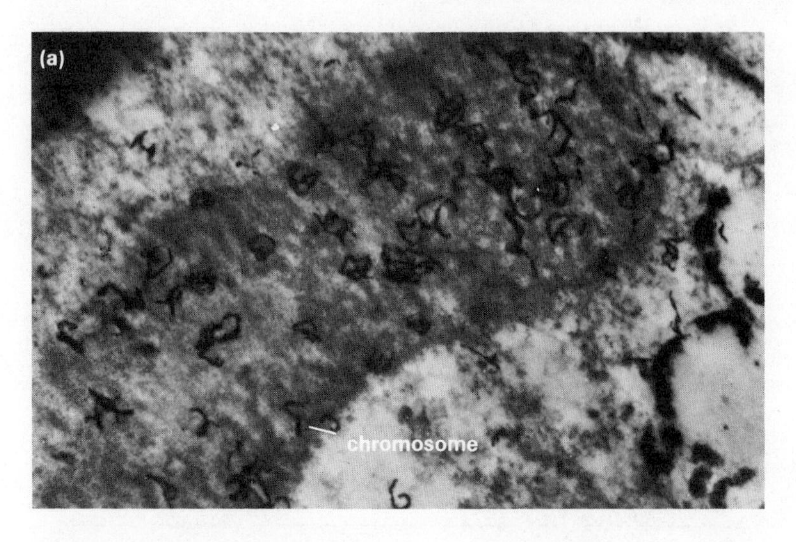

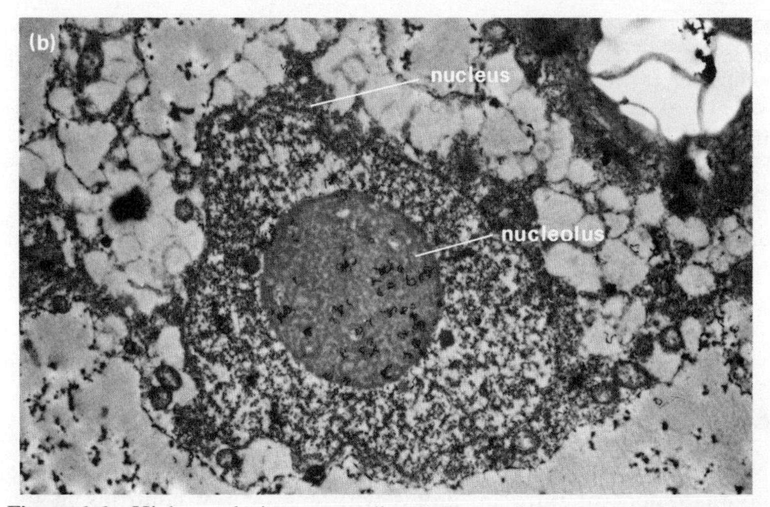

Figure 6.6 High resolution autoradiographs produced using labelled sections of *Spirogyra.* (a) Thymidine labelling of DNA in a section of a chromosome. (b) Uridine labelling of RNA in a nucleolus. In both cases the tritium label appears as black 'squiggles' on the photograph. (Photographs by courtesy of Dr E. G. Jordan, Biology Department, Queen Elizabeth College.)

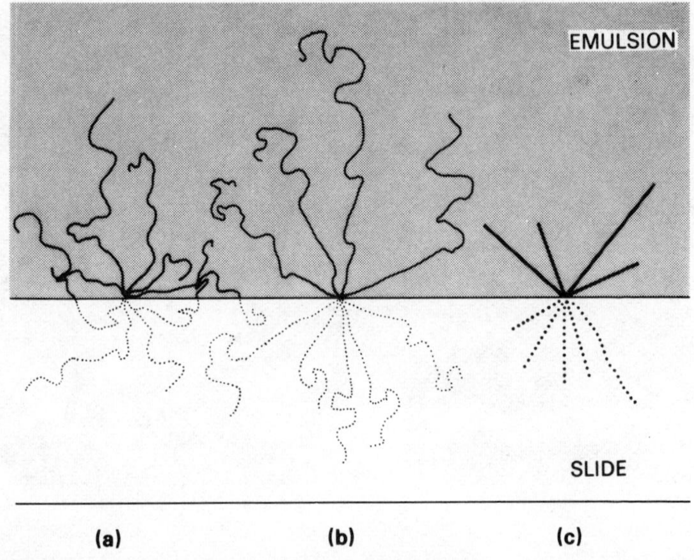

Figure 6.7 Representation of the distribution of α- and β-particle trajectories for labelled sources, mounted on a glass slide and covered with emulsion: (a) a medium energy β-source – particles cross the emulsion/slide interface freely in both directions; (b) a higher energy β-source – particles tend to have straighter trajectories nearer to the source showing less crossing over between emulsion and slide; (c) trajectory for an α-particle source. A very weak β-source such as tritium would produce very thin tracks since not many β-particles from such a source would be energetic enough to sensitize the emulsion. For very thin emulsions, particles from sources other than tritium either pass straight through without registering or produce a track at some distance from the source. For thin emulsions, therefore, tritium is the label of choice.

6.8 Radiocarbon dating

Naturally occurring ^{14}C is continually being created through the bombardment of nitrogen atoms with cosmic ray neutrons in the upper atmosphere:

$$^{14}_{7}N + ^{1}_{0}n \longrightarrow ^{14}_{6}C + ^{1}_{1}p$$

In this reaction the ^{14}N absorbs a neutron and emits a proton, thus changing to ^{14}C. The ^{14}C so produced is quickly oxidized to carbon dioxide which then equilibrates with the rest of the atmosphere. Therefore, when a plant is photosynthesizing it fixes a very small amount of ^{14}C along with the other carbon isotopes. In turn, all other living organisms can take up the ^{14}C as it passes through the food chain from plants. When the organism dies it ceases to take in any more ^{14}C, so that the level remaining in any residual tissues (e.g. wood, bone) depends only upon the initial amount present, the length of time elapsed since the organism died, and the rate of decay of the ^{14}C. Thus, in general terms, the longer an organism has been dead the less ^{14}C it contains.

The amount of $^{14}CO_2$ present in the atmosphere is approximately constant (overall, since the formation of the Earth, an equilibrium has been established between the rate of production of ^{14}C and its natural decay) but is subject to some fluctuation. In recent times thermonuclear weapons have increased the amount of ^{14}C in the atmosphere considerably (approximately a 90% increase in the Northern Hemisphere in 1963) so that present day air and organisms deriving carbon from it contain abnormally high amounts of ^{14}C and are hence useless as standards or for calibration values. In prehistoric times there are also thought to have been fluctuations in the $^{14}CO_2$ level due to changes in solar activity and in the Earth's magnetic field (both of which effect changes in the cosmic ray flux). Fortunately it has recently been possible to correlate these changing levels of $^{14}CO_2$ with absolute dates by comparing radiocarbon-derived dates with absolute dates obtained from tree rings, initially from the bristlecone pines (*Pinus aristata*), some living specimens of which are 4000 years old while other dead specimens in the same areas are as old as 8000 years (Fig. 6.8). As the figure shows, natural levels of $^{14}CO_2$ were higher in the past than now, so that initial levels of ^{14}C fixed were higher, and radiocarbon dating experiments must take this into account (higher initial levels give apparently younger ages!).

Two main methods are used in the actual process of dating – either liquid scintillation counting or gas counting. The carbon in the specimen (requiring perhaps 10–20 g of charcoal) is either converted to carbon dioxide or acetylene, which is then counted in a gas counter, or is converted to benzene for liquid scintillation counting. The background level in the counter, which must be very low, is determined using carbon from a very old sample of dead material (e.g. 300×10^6-year-old anthracite which contains no ^{14}C) and the counter is calibrated using a sample of known age (say, wood growing prior to 1850 to allow for modern fluctuations in atmospheric CO_2 levels), or an accepted artificial standard (a stock of oxalic acid is held by the US National Bureau of Standards for this purpose).

With these two determinations and accepting a value of 5730 a for the half-life of ^{14}C dating of samples up to 45 000 a can be attempted. In the future, new laser enrichment and mass spectroscopic techni-

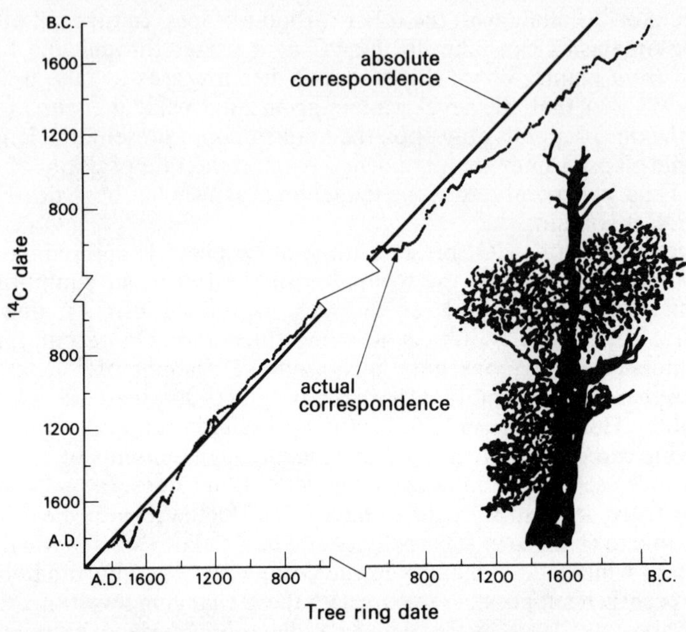

Figure 6.8 Actual correspondence between calendar date and ^{14}C-determined dates using ^{14}C age determination for samples of the bristlecone pine (*Pinus aristata*) together with the corresponding tree ring dates.

ques now becoming available may make it possible to measure samples as old as 100 000 a, with the added bonus of much smaller sample sizes.

6.9 Practical experiments

6.9.1 Counting dual labelled samples

Samples containing both carbon-14 and tritium can be counted in a liquid scintillation counter provided that the ratio of their activities falls within reasonable limits (see p. 114). Quench corrections can be applied by the channels ratio technique, but three separate channels are required (or two separate counts under changed conditions). On two-channel instruments it is usually more convenient to use the Automatic External Standard Channels Ratio (AESCR).

(a) Apparatus and materials

(1) Complete set of quenched standards separately containing carbon-14 and tritium

(2) Liquid scintillation counter
(3) Four 'unknown' samples containing a mixture of carbon-14 and tritium.

(b) Procedures Place all the samples in the counter in the order tritium standards first (unquenched to most quenched) followed by the carbon standards (unquenched to most quenched). Move the unquenched tritium sample into the counting chamber and switch the counter to REPEAT COUNT or MANUAL mode.

Set channel 1 to the optimum for carbon-14. Start the counter and slowly increase the lower discriminator until a negligible number (<0.1%) of tritium counts are observed. This channel is now set to count carbon-14 without interference from tritium.

With the tritium standard still in place set channel 2 to optimum for tritium and count the sample for 1 min. Next lower the setting of the upper discriminator until the count rate is reduced by approximately 10%. The purpose of this is to minimise the contribution of carbon-14 in the tritium channel. Now count all of the tritium and carbon standards using the AESCR method.

(c) Analysis of results Plot three quench correction curves on the same graph of E versus AESCR: curve 1 is for the efficiency of carbon-14 in channel 1; curve 2 is for the efficiency of carbon-14 in channel 2; curve 3 is for the efficiency of tritium in channel 2.

Use the AESCR values for the four 'unknowns' to read off from your graph the following: efficiency for carbon-14 in channel 1, E_1; efficiency for carbon-14 in channel 2, E_2; efficiency for tritium in channel 2, E_T.

Calculate the activities:

$$\text{Activity of carbon-14, } A_c = \frac{\text{net counts in channel 1}}{E_1}$$

$$\text{Tritium activity} = \frac{\text{net counts in channel 2} - A_c \times E_2}{E_T}$$

Record the tritium and carbon-14 activities for all four 'unknowns'.

Comment upon the accuracy of determination for the 'unknown' samples and draw conclusions about the applicability of the quench correction curves.

6.9.2 Determination of the amount of oxalic acid in rhubarb leaves by isotope dilution analysis

Rhubarb leaves contain relatively high amounts of oxalic acid together with other acids such as malic and citric acids. The oxalic acid, however, differs from the others in that it is only sparingly soluble in cold water and may, therefore, be purified by recrystallization from water.

(a) Apparatus and materials

(1) An aqueous solution of [^{14}C]oxalic acid containing approximately 0.05 g; 7.5 kBq (0.2 μCi) per cubic centimetre (the weight of oxalic acid should be known accurately)
(2) Rhubarb leaves (these can consist of both laminae and petioles)
(3) 500 ml beaker + clock glass lid
(4) Ion exchange column 20 cm long $\times$ 3 cm wide fitted with sintred glass disc at the bottom
(5) Anion exchange resin, Zeolit FFIP SRA 65 (standard resin, 14–52 mesh) or similar resin
(6) 4% (w/v) aqueous sodium chloride solution
(7) 2 mol dm^{-3} solution of hydrochloric acid
(8) Measuring cylinder (100 ml), filter paper and funnel, Pasteur pipettes and rubber teats, a selection of small filter beakers or 10 ml centrifuge tubes for recrystallization of small quantities of oxalic acid (the solubility of oxalic acid is $\simeq$10 g per 100 cm^3 of water at 20°C, and recrystallization of 200–500 mg quantities therefore requires less than 1 cm^3 of water).

(b) Procedure Weigh 100 g of shredded rhubarb leaves into the 500 ml beaker. Add exactly 1.00 cm^3 of the [^{14}C]oxalic acid solution followed by 50 cm^3 of distilled water, cover the beaker with the clock glass and boil the mixture for 5 min. Filter the mixture, retain the filtrate and re-extract the leaves with a further 35 cm^3 of boiling water for 5 min. Again filter the mixture, combine the filtrates and discard the extracted leaves into a low level active waste container.

Next prepare an ion exchange column of the anion exchange type. A suitable column can be made by introducing 35 cm^3 of Zeolit FFIP resin (14–52 mesh size) into a glass column (20 $\times$ 3 cm) and converting the resin into the chloride form by washing with 100 cm^3 of 4% NaCl solution followed by 100 cm^3 of distilled water (do not let the column run dry). Next run the filtrate through the column, when any oxalic acid will be retained on the resin. Wash the column with 50 cm^3 of distilled water and finally elute the oxalic acid off the column with dilute hydrochloric acid solution (2 mol dm^{-3}). Collect approximately 40 cm^3 of eluate.

Carefully evaporate the eluate to dryness in a round-bottomed or pear-shaped flask (a rotary evaporator is best, otherwise great care must be exercised to avoid splashing and thermal decomposition of the product. An alternative procedure is to partly immerse the flask in a heated bath at 60–70°C and direct a gentle stream of nitrogen through a Pasteur pipette at the surface of the solution until all the water has evaporated). Dissolve the product in the minimum amount of boiling distilled water (usually 1 cm^3 or less) and using a heated Pasteur pipette transfer the hot solution to a small recrystallization beaker (or tapered centrifuge tube). Cool the beaker in ice until the oxalic acid has crystallized, then remove the mother liquors and recrystallize a second time. At this point the oxalic acid should be dried in a desiccator and a small portion of the dry crystals accurately weighed and used to determine the specific activity, S (see below).

The remainder of the crystals should be recrystallized, dried and again a small portion used to determine S. This process is repeated until a constant value for S is obtained (within the limits set by counting and weighing errors).

The specific activity, S, is best determined by dissolving the crystals in water then mixing with a suitable liquid scintillant (see p. 63) and counting in a liquid scintillation counter. Similarly the specific activity of the original [^{14}C]oxalic acid, S_0, may be determined by counting aliquots of the stock solution. The amount of oxalic acid originally present in the leaves may then be calculated using equation 6.1 (p. 102).

Appendix A

Some useful isotopes

Element	*Atomic weight*	*Radio-isotope*	*Half-life*	*Decay mode*	*Energy of radiation (MeV)*	
					β (max.)	γ
hydrogen	1.008	^{3}H	12.26 a	β^-	0.018	—
carbon	12.01	^{14}C	5730 a	β^-	0.156	—
sodium	22.99	^{22}Na	2.6 a	β^+ (90.5%)	0.54	1.28
				EC (9.5%)	—	0.51 (ann.)
		^{24}Na	15.0 h	β^-	1.39	1.37, 2.75
phosphorus	30.97	^{32}P	14.3 d	β^-	1.71	—
sulphur	32.06	^{35}S	87.4 d	β^-	0.167	—
chlorine	35.45	^{36}Cl	3×10^5 a	β^- (98.3%)	0.714	—
				EC (1.7%)	—	
potassium	39.10	^{42}K	12.4 h	β^-	2.0 (18%) 3.6 (82%)	1.52 (18%)

calcium	40.08	^{45}Ca	165 d	β^-	0.254	—
chromium	52.00	^{51}Cr	27.8 d	EC	—	0.323 (9%)
manganese	54.94	^{54}Mn	314 d	EC	—	0.84
iron	55.85	^{55}Fe	2.7 a	EC	—	0.0059 (23%)
		^{59}Fe	45 d	β^-	0.27 (46%)	1.10 (57%)
					0.46 (54%)	0.19 (2.4%)
cobalt	58.93	^{57}Co	270 d	EC	—	0.136 (8.8%)
						0.122 (88.8%),
						0.014 (8%)
		^{60}Co	5.26 a	β^-	0.31	1.17, 1.33
copper	63.55	^{64}Cu	12.8 h	β^- (38%)	0.57	1.34 (0.6%)
				β^+ (19%)	0.66	0.51 (ann.)
				EC (43%)	—	
zinc	65.38	^{65}Zn	245 d	β^+ (1.7%)	0.325	1.11 (49%)
				EC (98.3%)	—	0.51 (ann.)
bromine	79.90	^{82}Br	35.4 h	β^-	0.44	many in range 0.55–1.47
strontium	87.62	^{90}Sr	28 a	β^-	0.54	—
					(2.27, ^{90}Y daughter)	
tin	118.69	^{113}Sn	119 d	EC	—	0.26 (2%)
iodine	126.91	^{125}I	60 d	EC	—	0.035 (7%)
		^{131}I	8.04 d	β^-	0.81 (1%)	0.72 (2.8%), 0.64 (9.3%)
					0.61 (86%)	0.36 (81%), 0.28 (5%)
					0.34 (13%)	
caesium	132.91	^{137}Cs	30 a	β^-	0.52 (92%)	0.66 (82%)
gold	196.97	^{198}Au	2.7 d	β^-	0.29 (1.2%)	1.09 (0.2%), 0.7 (1%)
					0.86 (98.8%)	0.412 (95.8%)

Appendix B

Radiation protection data

Radio-isotope	*Toxicity class*†	*Annual limit of intake*‡ *(Bq)*		*γ-dose rate at 1 m* § *(μGyh*$^{-1}$ MBq^{-1}*)*	*Maximum activity per student experiment*‖	
		Oral	*Inhalation*		*(Bq)*	*(μCi) (approx.)*
^{3}H	4	3×10^{9}¶	3×10^{9}¶	—	3×10^{8}¶	8×10^{3}¶
^{14}C	3	(2×10^{8})	(4×10^{8})	—	2×10^{7}	500
^{22}Na	2	2×10^{7}	2×10^{7}	0.30	2×10^{6}	50
^{24}Na	3	1×10^{8}	2×10^{8}	0.47	1×10^{7}	250
^{32}P	3	2×10^{7}	1×10^{7}	—	1×10^{6}	25
^{35}S	3	2×10^{8}	8×10^{7}	—	8×10^{6}	200
^{36}Cl	2	6×10^{7}	9×10^{6}	—	9×10^{5}	25
^{42}K	3	2×10^{8}	2×10^{8}	0.034	2×10^{7}	500
^{45}Ca	2	6×10^{7}	3×10^{7}	—	3×10^{6}	80
^{51}Cr	3	1×10^{9}	7×10^{8}	0.004	7×10^{6}	150
^{54}Mn	2	7×10^{7}	3×10^{7}	0.118	3×10^{6}	80
^{55}Fe	3	3×10^{8}	7×10^{7}	—	7×10^{6}	150

^{59}Fe	3	3×10^7	1×10^7	0.160	1×10^6	25
^{57}Co	3	2×10^8	2×10^7	0.024	2×10^6	50
^{60}Co	2	7×10^6	1×10^6	0.334	1×10^5	2.5
^{64}Cu	3	4×10^8	8×10^8	0.030	4×10^7	1000
^{65}Zn	3	1×10^7	1×10^7	0.078	1×10^6	25
^{82}Br	3	1×10^8	1×10^8	0.367	1×10^7	250
^{90}Sr	1	1×10^6	1×10^5	—	1×10^4	0.25
^{113}Sn	3	(2×10^7)	(4×10^7)	0.067	2×10^6	50
^{125}I	2	1×10^6	2×10^6	0.034	1×10^5	2.5
^{131}I	2	1×10^6	2×10^6	0.051	1×10^5	2.5
^{137}Cs	2	4×10^6	6×10^6	0.083	4×10^5	10
^{198}Au	3	4×10^7	4×10^7	0.061	4×10^6	100

†Taken from *A basic toxicity classification of radionuclides.* Vienna: International Atomic Energy Agency, 1963.

‡Taken from *Limits for intakes of radionuclides by workers*, ICRP Publication 30, Part 1, *Annals of the ICRP* **2**(3/4), 1979; Part 2, *Annals of the ICRP* **4** (3/4), 1980. Figures in brackets calculated from the data in *Handling, storage, use and disposal of unsealed radionuclides in hospitals and medical research establishments*, ICRP Publication 25, *Annals of the ICRP*, **1**(2), 1977: these figures are subject to reappraisal by the ICRP and revised recommendations are to be published.

§Calculated from data in ICRP Publication 25 (see above).

‖Calculated according to the recommendations in *Radiation protection in schools*. ICRP Publication 13. Oxford: Pergamon Press, 1970.

¶This value is for tritiated water. For organic compounds such as tritiated thymidine the value should be reduced to one fiftieth of the quoted value.

Appendix C

Units, prefixes, derived units and constants

Basic SI units

Physical quantity	*Unit*	*Symbol*
length	metre	m
mass	kilogram	kg
time	second	s
electric current	ampere	A
amount of substance	mole	mol

Prefixes

Multiple	*Prefix*	*Symbol*
10^{12}	tera	T
10^{9}	giga	G
10^{6}	mega	M
10^{3}	kilo	k
10^{2}	hecto	h
10	deca	da
10^{-1}	deci	d
10^{-2}	centi	c
10^{-3}	milli	m
10^{-6}	micro	μ
10^{-9}	nano	n
10^{-12}	pico	p

Derived SI units

Physical quantity	*Unit*	*Symbol*	*Definition*
area	square metre	m^2	—
volume	cubic metre	m^3	—
concentration	mole per cubic metre	$mol\ m^{-3}$	—
density	kilogram per cubic metre	$kg\ m^{-3}$	—

frequency	hertz	Hz	s^{-1}
energy, enthalpy	joule	J	$kg\ m^2\ s^{-2}$
force	newton	N	$J\ m^{-1}$
power	watt	W	$J\ s^{-1}$
electric potential difference	volt	V	$J\ A^{-1}\ s^{-1}$
activity	becquerel	Bq	s^{-1}
absorbed dose	gray	Gy	$J\ kg^{-1}$
dose equivalent	sievert	Sv	$J\ kg^{-1}$

Constants

Physical constant	*Symbol*	*Value*
speed of light	c	$2.997\ 925 \times 10^8\ m\ s^{-1}$
unified atomic mass constant (one-twelfth of mass of a ^{12}C atom)	u	$1.660\ 43 \times 10^{-27}\ kg$
mass of proton	m_p	$1.6726 \times 10^{-27}\ kg$
mass of neutron	m_n	$1.6749 \times 10^{-27}\ kg$
mass of electron	m_e	$9.1095 \times 10^{-31}\ kg$
Planck's constant	h	$6.624\ 91 \times 10^{-34}\ J\ s$

Non-SI units and conversion factors

Physical quantity	*Unit*	*Symbol*	*Conversion factor*
length	inch	in	$2.54 \times 10^{-2}\ m$
energy	calorie	cal	4.184 J
energy	electronvolt	eV	$1.602 \times 10^{-19}\ J$
activity	curie	Ci	$3.7 \times 10^{10}\ Bq$
absorbed dose	rad	rad	0.01 Gy
dose equivalent	rem	rem	0.01 Sv

Appendix D

A selected bibliography for further reading

Autoradiography

Fisher, H. A. and M. R. F. Ashworth 1975. *An introduction to practical autoradiography*. Review no. 15. Amersham: The Radiochemical Centre.

Gahan, P. B. (ed.) 1972. *Autoradiography for biologists*. New York and London: Academic Press.

Rogers, A. W. 1980. *Techniques of autoradiography*, 4th edn. Amsterdam: Elsevier.

General

Catch, J. R. 1961. *Carbon-14 compounds*. London: Butterworths.

Coomber, D. I. 1975. *Radiochemical methods in analysis*. London: Plenum Press.

Evans, E. A. 1974. *Tritium and its compounds*, 2nd edn. London: Butterworths.

Faires, R. A. and G. G. J. Boswell 1981. *Radioisotope laboratory techniques*, 4th edn. London: Butterworths.

Glasstone, S. 1967. *Sourcebook on atomic energy*, 3rd edn. New York: Van Nostrand, Reinhold.

Kamen, M. D. 1957. *Isotopic tracers in biology*. New York: Academic Press.

McKay, H. A. C. 1971. *Principles of radiochemistry*. London: Butterworths.

Raaen, V. F., G. A. Ropp and H. P. Raaen 1968. *Carbon-14*. New York: McGraw-Hill.

Wang, C. H., D. L. Willis and W. D. Loveland 1975. *Radiotracer methodology in the biological, environmental and physical sciences*. Englewood Cliffs, NJ: Prentice-Hall.

Isotope dilution analysis

Anon. 1965. *Radioactive isotope dilution analysis*. Review no. 2. Amersham: The Radiochemical Centre.

Ayrey, G., D. Barnard and T. R. Houseman 1971. Use of radioisotopically labelled analytical reagents in organic chemistry. *Chem. Rev.* **71**, 371.

Tolgyessy, J., T. Braun and M. Krys 1972. *Isotope dilution analysis*. Oxford: Pergamon Press.

Isotope effects

Melander, L. and W. H. Saunders 1980. *Reaction rates of isotopic molecules*. New York: John Wiley.

Thomson, J. F. 1963. *Biological effects of deuterium.* Oxford: Pergamon Press.

Liquid scintillation counting

Birks, J. B. 1964. *The theory and practice of scintillation counting.* Oxford: Pergamon Press.

Crook, M. A. and P. Johnson (eds) 1970–77. *Liquid scintillation counting*, vols 1–5. London: Heyden.

Dyer, A. 1980. *Liquid scintillation counting practice.* London: Heyden.

Fox, B. W. 1976. *Techniques of sample preparation for liquid scintillation counting.* Amsterdam: North-Holland.

Horrocks, D. L. 1974. *Applications of liquid scintillation counting.* New York: Academic Press.

Kobayashi, Y. and D. V. Maudsley 1974. *Biological applications of liquid scintillation counting.* New York: Academic Press.

Neame, K. D. and C. A. Homewood 1974. *Introduction to liquid scintillation counting*. London: Butterworths.

Radioactivation analysis

Kruger, P. 1971. *Principles of activation analysis*. New York: John Wiley.

Taylor, D. 1964. *Neutron irradiation and activation analysis.* London: Newnes.

Radioimmunoassay

Bolton, A. E. 1977. *Radioiodination techniques.* Review no. 18. Amersham: The Radiochemical Centre.

Freeman, M. L. and M. D. Blaufox (eds) 1975. *Radioimmunoassay*. New York: Grune & Stratton.

Hunter, W. M. 1973. Radioimmunoassay. In *Handbook of experimental immunology*, D. M. Weir (ed.), Ch. 17. Oxford: Blackwell Scientific.

Radiation protection

Anon. 1970. *Radiation protection in schools for pupils up to the age of 18 years.* ICRP Publication 13. Oxford: Pergamon Press.

Anon. 1971. *Handbook of radiological protection, Part I: Data.* London: HMSO.

Anon. 1963. *A basic toxicity classification of radionuclides.* Technical Report Series no. 15. Vienna: International Atomic Energy Agency.

Anon. 1977. *Recommendations of the International Commission on Radiological Protection.* ICRP Publication 26. *Annals of the ICRP* **1**(3).

Anon. 1977. *The handling, storage, use and disposal of unsealed radionuclides in hospitals and medical research establishments.* ICRP Publication 25. *Annals of the ICRP* **1**(2).

Health and Safety Executive 1977. *Guidance notes for the protection of persons exposed to ionising radiations in research and teaching.* London: HMSO.

Martin, A. and S. A. Harbison 1980. *An introduction to radiation protection.* 2nd edn. London: Chapman & Hall.

Wrixon, A. D., G. S. Linsley, K. C. Binns and D. F. White 1979. *Derived limits for surface contamination.* NRPB-DL2. Harwell: National Radiological Protection Board.

Storage and stability

Bayly, R. J. and E. A. Evans 1968. *Storage and stability of compounds labelled with radioisotopes.* Review no. 7. Amersham: The Radiochemical Centre.

Evans, E. A. 1976. *Self-decomposition of radiochemicals*. Review no. 16. Amersham: The Radiochemical Centre.

Appendix E

Some useful addresses

For instruments

AID Scientific Ltd	Canford Works Wallisdown Road Bournemouth Dorset BH11 8QW UK	
Baird Corp.	125 Middlesex Tnpk. Bedford MA 01730 USA	Baird-Atomic Ltd. Warner Drive Rayne Road Braintree Essex UK
Beckman Instruments Inc.	2500 Harbor Blvd. Fullerton CA 92634 USA	Beckman RIIC Ltd. Turnpike Road High Wycombe Bucks HP12 3NR UK
Canberra Industries	45 Gracey Avenue Meriden CT 06450 USA	5 Pioneer Road Faringdon Oxon SN7 7BU UK
ESI Nuclear	See AID Scientific Ltd.	
Kontron Ltd	Bernerstrasse-Sued 169 Zurich Switzerland 8048	PO Box 88 St. Albans Herts, AL1 5JG UK
LKB-Produkter AB	Fack Bromma Sweden S-161 25	LKB House 232 Addington Road Selsdon South Croydon Surrey CR2 9PX UK
Nuclear Data Inc.	Golf & Meacham Rds. Schaumburg IL 60196 USA	Rose Industrial Estate Cores End Road Bourne End Bucks UK
Nuclear Enterprises Ltd	Sighthill Edinburgh EH11 4EY UK	

Packard Instrument Co.	2000 Warrenville Road Downers Grove IL 60515 USA	13–17 Church Road Caversham Berks RG4 7AA UK
Panax Nucleonics Ltd	See AID Scientific Ltd	
N.V. Philips (Gloeilampen fabricken)	Building TQ 111-2 Eindhoven The Netherlands	Pye-Unicam Philips Analytical Dept. York Street Cambridge CB1 2PX UK
Searle Analytic Inc.	2000 Nuclear Dr. Des Plaines IL 60018 USA	Lane End Road Sands High Wycombe Bucks

For chemicals and scintillation chemicals

BDH Chemicals Ltd	Poole Dorset BH12 4NN UK	
Eastman Kodak Co.	343 State St. Rochester NY 14650 USA	Kirkby Liverpool UK
Fisons Scientific Apparatus	Bishop Meadow Road Loughborough Leicestershire LE11 OR4 UK	
Hopkins and Williams Ltd	PO Box 1 Romford Essex RM1 1HA UK	
Koch-Light Laboratories Ltd	2 Willow Road Colnbrook Bucks SL3 0BZ UK	
E. Merck	Frankfurter Str. 250 Postfach 4119 Darmstadt W. Germany D-6100	See also BDH Chemicals

For radiochemicals

Amersham International Ltd (The Radiochemical Centre)	White Lion Road Amersham Bucks UK	Amersham Corp. 2636 S. Clearbrook Dr. Arlington Heights IL 60005 USA

ICN Chemical & Radioisotope Division	2727 Campus Dr. Irvine CA 92715 USA	
Imperial Chemical Industries Ltd	Physics & Radioisotope Services PO Box 2 Billingham Cleveland TS23 1JB UK	
New England Nuclear	549 Albany St. Boston MA 08817 USA	2 New Road, Southampton SO2 0AA UK

Glossary

activation The production of a radioactive isotope by bombardment of a stable isotope with nuclear projectiles such as neutrons, protons or α-particles.

activity The rate of disintegration of a radioactive nuclide. Units of activity are the becquerel ($1\ Bq = 1\ s^{-1}$) and the curie ($1\ Ci = 3.7 \times 10^{10}\ s^{-1}$).

alpha particle (α-particle) The nucleus of a helium atom (2 protons + 2 neutrons) emitted by radioactive elements with $Z > 82$.

annihilation radiation Two γ-rays of 0.51 MeV energy resulting from the mutual destruction of a positron and an electron.

anthracite A hard mineral coal, burning without smoke and giving intense heat.

antibody A globular protein produced within an animal which combines with an antigen and renders it harmless.

antigen Any foreign substance, generally protein, which causes the formation of specific antibodies within an animal.

antisera The clear fluid left after coagulation and removal of all the cells from an animal's blood and containing one or more antibodies produced as a result of previous exposure of the animal to one or more antigens.

atomic mass The weight of an atom measured in unified atomic mass units (1u = 1/12 of mass of a carbon-12 atom).

atomic number, Z The number of protons in the nucleus of a given element.

auger electrons An electron ejected from an atom with energy released by another electron filling a vacancy in one of the inner electron shells.

axon A process of a nerve cell which carries the nerve impulse away from the nerve cell body.

beta particle (β-particle) A negatively or positively charged electron which is emitted by a radioactive atom leaving either an extra proton (β^- decay) or neutron (β^+ decay) in the nucleus.

brehmsstrahlung Electromagnetic radiation, similar to γ-radiation, which is emitted when charged particles are retarded as they pass close to a positively charged nucleus.

carrier A non-radioactive element or compound which is present or is deliberately added to its radioactive counterpart thus reducing the specific activity.

carrier free A radioactive species which is undiluted by the corresponding stable isotopic species.

cartilage A connective tissue which helps to support an organism and hold its various parts together.

channel The 'window' in a pulse height analyser which is defined by an upper and a lower energy level set by means of **discriminators**.

coincidence circuit An electronic device capable of deciding whether or not pulses from two radiation detectors resulted from the same radioactive event.

collagen A type of fibrous protein found in connective tissues.
counting efficiency The ratio of the observed net count-rate of an assay sample to the absolute disintegration rate.
curie (Ci) Unit of activity defined as $3.7 \times 10^{10}\ s^{-1}$.
cyclotron A machine for accelerating positive ions to high energies so as to promote nuclear reactions.
dead time The period, following detection of an ionising particle or ray, when a radiation detector is unable to detect further radiations.
discriminator An electronic device for rejecting or accepting electrical pulses of any given size.
dose The amount of energy absorbed by tissue or other material subject to radiation.
dosemeter A device for measuring the amount of radiation absorbed.
electron A stable elementary particle with unit negative charge weighing approximately 1/1800 of a proton.
electron capture A mode of radioactive decay in which an electron from an inner shell is captured by a proton in the nucleus to form a new nuclide with one less proton ($Z - 1$) and one more neutron ($N + 1$).
excitation Occurs when a nucleus, atom or molecule absorbs energy and changes from the ground state to an excited state.
fission A nuclear reaction in which a large nucleus (e.g. uranium-235) splits into two roughly equal parts releasing a large amount of energy.
fission products The products resulting from a nuclear fission (at least 250 are known). They are the obnoxious constituents of nuclear bomb 'fallout' and possible environmental contaminants from nuclear power stations.
gamma radiation (γ-radiation) Short wavelength electromagnetic radiation emitted by excited nuclei as they return to a lower energy state.
ground state The lowest energy quantum state of a nucleus, atom or molecule.
half-life The time taken for the activity of a radioactive isotope to decay from a given value to exactly half that value.
hormone A substance produced in small quantities in one part of an organism which diffuses or is transported to another part of the organism where it profoundly affects metabolism. Hormones can be steroids (e.g. oestrogens), proteins (e.g. insulin) or small metabolites (e.g. adenosine-3′, 5′-cyclic monophosphate).
internal conversion (IC) Ejection of an electron from an atom with energy resulting from an excited nucleus undergoing a transition to a lower energy state. The process competes with γ-ray emission.
isomer (nuclear) A nucleus in an excited state with a finite half-life.
isotopes A series of nuclides with the same atomic number (i.e. same number of protons) but differing numbers of neutrons.
mass number, A The sum of the number of protons, Z, and the number of neutrons, N, in a nucleus. $A = Z + N$.
narcotic A substance which induces sleep and in large doses can produce insensibility and stupor.
neutron An uncharged elementary particle of mass $\simeq 1u$. An important constituent of atomic nuclei.

neutrino A tiny particle with no charge and a negligibly small mass.

nuclear projectile Any high energy particle or photon which is used to bombard atomic nuclei.

nuclear reaction Interaction between a particle or photon and an atomic nucleus which results in the formation of a new atomic species.

nuclear reactor An arrangement of fissionable material (e.g. uranium-235) inside a neutron reflecting material or 'moderator' such as graphite or deuterium oxide together with neutron absorbing material such as cadmium or boron in the form of control rods. By careful adjustment of the geometry of the three components a controlled self sustaining fission chain reaction can be maintained which liberates enormous quantities of energy for generation or electricity or motive power.

nucleic acids Biological macromolecules consisting of polymers of nucleotides responsible for the storage translation and expression of genetic information. (Nucleotides are themselves composed of phosphate, a pentose sugar, and a purine or pyrimidine base.)

nucleon The collective term for the two major particles (proton and neutron) which are found in the nucleus.

nucleus The dense central core of an atom which carries nearly all the mass and is composed of protons and neutrons.

nuclide An atomic species with a specified nucleon content.

pair production The conversion of energy into mass in the form of a positron and an electron which can occur when a γ-photon with energy greater than 1.02 MeV passes near a nucleus or an electron.

photon A quantum of light or electromagnetic energy which has properties similar to a particle.

positron A positively charged electron which is part of 'antimatter'.

proton A stable elementary particle with unit positive charge, the nucleus of a hydrogen atom.

radioisotope A nuclide whose nucleus is unstable and eventually undergoes radioactive decay.

resolving time See **dead time**.

reticulocyte An immature red blood cell with no nucleus but containing a net-like reticulate cytoplasm.

scintillant, scintillator A material which emits flashes of light when excited by ionizing radiation.

self decomposition The production of impurities in a radiochemical due to degradation of the pure compound caused by its own ionizing radiations.

solubilizer A substance which will dissolve tissue or other material and enable it to be mixed with a liquid scintillant solution.

specific activity The ratio of the number of radioactive atoms to the total number of atoms of the same element in a given compound. Often more loosely defined as activity per unit weight.

steroid A group of complex lipid-type molecules having a characteristic poly-cyclic carbon nucleus containing 17 carbon atoms (includes sterols, sex hormones and vitamin D).

tracer An isotopic species (stable or radioactive) introduced into a chemical reaction or biological system in order to trace the path normally taken by the abundant stable species of the same element, e.g. radioactive carbon-14 is used as a tracer for the element carbon whose most abundant species is carbon-12.

X-ray Electromagnetic radiation which arises when fast electrons strike a solid target. They consist partly of *brehmsstrahlung* and partly of radiation resulting from electron movements in the inner shells of the target atoms.

Index

Numbers printed in italics refer to text figures